SpringerBriefs in Applied Sciences and Technology

PoliMI SpringerBriefs

AF900264

More information about this subseries at http://www.springer.com/series/11159
http://www.polimi.it

Anna Dalla Valle

Change Management Towards Life Cycle AE(C) Practice

POLITECNICO
MILANO 1863

Springer

Anna Dalla Valle
Department of ABC
Politecnico di Milano
Milan, Italy

ISSN 2191-530X ISSN 2191-5318 (electronic)
SpringerBriefs in Applied Sciences and Technology
ISSN 2282-2577 ISSN 2282-2585 (electronic)
PoliMI SpringerBriefs
ISBN 978-3-030-69980-2 ISBN 978-3-030-69981-9 (eBook)
https://doi.org/10.1007/978-3-030-69981-9

This Springer imprint is published by the registered company Springer Nature Switzerland AG
The registered company address is: Gewerbestrasse 11, 6330 Cham, Switzerland

Foreword by Colin Rohlfing

To date, there is an established consensus in the design and construction industry about the importance of Life Cycle Thinking (LCT) to move the built environment towards a more regenerative future. However, this consensus does not always translate into practice given perceived project constraints such as budget, schedule, scope and even ignorance. It is important to list them as "perceived" since their true identities are not "constraints", rather "opportunity" and "design parameters", both of which designers have a love of addressing head-on. One great opportunity is the use of LCT in design thinking to address the fact that even Zero Net Energy (ZNE) buildings take 10–15 years to overcome the environmental price of their making due to the embodied emissions associated with building materials. Given the fact that we will reach an irreversible tipping point for the impacts of climate change in 10 years' time, the world does not have the luxury to wait that long.

To fulfil the potential of LCT and related methodologies, a systemic change is needed in the way products, services, systems, infrastructures and buildings are designed and maintained. For designers, this change imposes a range of new opportunities and requires novel approaches. To develop products, buildings, and services with a life cycle perspective, the entire life cycle including production, construction, use and end-of-life phases needs to be addressed from the day of the design phase. This requires architects and engineers to assume a holistic and systematic approach during design, focusing to a much greater extent on the temporal aspects of how a building or product might function and change over time.

Due to the complexities of the construction sector, the wide range of building requirements and the plurality of diverse players in the supply chain, the integration of life cycle thinking in the whole building process demands a shift in both mindset and project process, as clearly evidenced by the book.

In this context, in addition to exploring in depth the available methods and tools to support life cycle design, the author reports in detail on how environmental issues and life cycle perspective are being implemented in practice by various design practitioners. The presented critical analysis of current design practice in terms of involved actors, performed studies, tools and considered information turns out to be pivotal for the improvement and optimization of the design process towards environmental sustainability. Moreover, the proposed assessment framework with the inferred worksheets makes a compelling argument for switching to the expected life cycle AE(C) practice, providing an overall vision of the required workflow at the different process phases.

What is particularly interesting in the book is the ethnographic approach adopted for filling the gap between theory and practice in the design and construction field. Understanding the customs and culture created within the field—and more importantly, within design practice—is the key to a true shift in human behaviour and design thinking. I often tell people that the role of a Sustainable Design Leader is not the mastery of technical matters, but the understanding of human psychology and behaviour change within the context of an organization. The importance of this understanding is highlighted within the book as it suggests the need of sharing responsibilities among design practitioners, empowering the different disciplines in the integration of LCT into the design process. In addition, the resulting project-based life cycle database envisioned in a BIM-oriented working environment plays a key role not only to provide new competitive advantages for design firms and added value for clients but also to provide a long-term vision of the built environment.

This vision is one of the Carbon-Balanced buildings, which are created with components that sequester more carbon over their lifecycle than they emit. This is a true Zero Net Carbon reality that must consider buildings as carbon banks throughout its entire life cycle. This reality requires change management from current practice to life cycle AE(C) practice and the involvement in the decision-making process of the suggested new intermediary between clients and project team to face the complexity of the building sector and to support the change of mindset and in the way of practice towards life cycle design.

Chicago, USA

Colin Rohlfing, LEED AP BD+C
Director of Sustainable Development, HDR

Foreword by Andrea Campioli and Monica Lavagna

Three closely correlated issues are radically changing the way we conceive design activity in the built environment transformation processes.

First, the climate change mitigation policy requires unprecedented effort to identify methodologies, techniques and tools that allow a considerable reduction of the environmental impacts. The construction industry plays a crucial role in the mitigation of climate change, representing 39% of annual energy-related CO_2 emissions at a global level. Moreover, it constitutes a growing sector, since the overall floor area of buildings is expected to double by 2060. To date, the existing policy commitments appear to be insufficient, calling for additional actions (UNE 2017).

Second, digitalization has made available ever more sophisticated and powerful instruments, conceived as tools no longer simply to build but to think (Bernstein 2018), taking full advantage of data and information to simulate future scenarios.

Third, the complexity between the challenges to be faced and the tools to solve them requires increasingly specialized expertise that can be involved only in medium- and big-size firms. In the fast-moving market, the capacity of design firms to handle organizational change is needed for maintaining competitiveness (Lines and Reddy Vardireddy 2017).

In this context, on the one hand, life cycle perspective and related methodologies, especially Life Cycle Assessment (LCA) and Life Cycle Costing (LCC), are assuming an emerging role, and on the other, Architectural, Engineering and Construction (AEC) firms are pursuing innovation efforts in terms of both tangible (e.g. materials, equipment and tools) and intangible resources (e.g. knowledge, management and competences).

As regards the change management of medium- and big-size design firms, the book explores an extremely topical issue: the urgent need to integrate environmental efficiency variables into the decision-making process.

It is important to emphasize that here design firms have been surveyed taking ethnography and related tools as a reference methodology. A two-level analysis, typical of ethnographic studies, has been performed for dealing with the change management induced in the construction industry by the paradigm spread of environmental sustainability. On the one hand, the leading AEC firms established at the national and international levels have been interviewed, and on the other, a sample of case studies has been deepened by the author, through a personal involvement in some design experiences of one of the largest A/E firms in the world, with the aim to map the non-linear path of the different building projects from the concept phase to the construction document phase. This methodological approach is based on direct contact with both architects and engineers, looking at their backgrounds and daily work and the workflows among practitioners by asking questions and reporting in detail (O'Reilly 2012) the design activities. In this way, the book examines extensively and well-documents the design and construction fields, highlighting the complex relationship between the different actors of the design process in handling environmental issues and the crucial role of adequate information according to the different project phases of development.

From practitioners' point of view, Building Information Modelling (BIM) is increasingly recognized as a reference tool to adequately support decision-making in the effort of integrating environmental issues into the design process, by systematizing several design aspects and managing a large amount of information. The book offers a broad overview of the available LCA tools and BIM-based LCA methods, describes in detail the information flows and the practitioners' relationships of current highly-complex design practice and identifies the possible ways of managing environmental information in a life cycle perspective along the entire decision chain.

In addition, in order to support the transition towards life cycle design and operations, an assessment framework is proposed to implement Life Cycle Thinking into the design process, revealing the life cycle information required at the different phases and empowering different actors. The existing information flow is showed in order to point out how current practice works nowadays to meet environmental and life cycle issues, disclosing in particular the core competences and related networks. Due to the wide range of information demanded to switch from current design practice to a life cycle AE(C) practice, a sharing of responsibilities is aimed, as evidenced by the proposed worksheets intended to assign to each competence the committed technological elements and connected life cycle information. The goal is encouraging a progressive implementation of information targeting a growing level of detail and accuracy of data over the process, encouraging the team in adopting a systemic approach to life cycle design and in developing a project-based and well-framed life cycle database (envisioned in a BIM-oriented working environment) for providing a long-term vision of the facility and a continuous build-up of know-how. The resulting change management calls for, especially in particularly complex projects, a new professional figure that, recalling the "middleware", acts as an intermediary between clients and the project team towards life cycle design.

In conclusion, the book offers a broad and extensive vision of the skills and tools currently exploited by design firms for managing life cycle issues, in particular the environmental ones, in the decision-making processes, identifying key roles and levels of mutual interaction.

The result is an original reshaping of the information flows in relation to the different design phases and to the miscellaneous competence involved, as well as some future outlook.

Andrea Campioli
Monica Lavagna
Department of ABC
Politecnico di Milano
Milan, Italy

References

Bernstein PG (2018) Architecture, Design data. Practice Competency in the era of computation. Birkhäuser: Basel

Lines BC, Reddy Vardireddy PK (2017) Drivers of organizational change within the AEC industry: linking change management practices with successful change adoption. Journal of Management in Engineering 33 (6): 04017031

O'Reilly K (2012) Etnographic Methods. Routledge: New York

United Nations Environment, International Energy Agency (2017) Towards a zero-emission, efficient, and resilient buildings and construction sector. Global status report 2017

Preface

The book takes place in the ongoing process of transformation and increasing complexities of the building sector, characterized by the globalization of the market, the broad range of requirements, the specialization of competencies, the wide set of tools and the digitalization of the process. Here, environmental issues are gaining even more attention, becoming part of many international agenda, standards and codes, bringing out new pressure and more radical changes. The transformation had a great impact to the point that, while until a short time ago environmental targets were seen as constraints (e.g. energy efficiency directive), today they are even more considered as a way to improve performance and increase competitiveness. The result is that, nowadays, more or less every design firm claims to be environmentally friendly to take advantage of their business.

In this context, the aim is to understand how Architecture, Engineering and Construction (AEC) firms are equipping and reorganizing themselves in order to address and meet environmental issues. From a theoretical point of view, several studies were performed in the scientific community dealing with sustainability and environmental aspects in the construction sector. On the one hand, they analysed advancements, strengths, weaknesses and all the connected topics to clarify the current state and, on the other, they explored new methodologies, frameworks, instruments and tools to support design tasks. By contrast, from a practical point of view, only few studies were developed to figure out what really happens in design and construction practices. Concerning this, many are yet the open questions to be solved: the environmental issues taken into consideration, their role within the decision-making process, the adopted way of practice, the used set of tools and software, the involved skills and competences and the related information flow.

Since this field remains until now blurry, the book seeks to bridge the gap, focusing on AEC practice and identifying all the tangible and intangible resources invested by firms to achieve environmental goals and their influence within the decision-making process. In particular, three entities are taken into consideration: (i) the team of actors, including all practitioners and experts involved in the design process; (ii) the set of tools and the assets, including all physical items, computer and software necessary to design; and (iii) the collection of data, including all information required both by experts and tools to work and design. Moreover, since there is no pre-determined relationship between the resources of a firm and its capabilities, the purpose is to figure out how they are linked together and the related information flow within the decision-making process.

Indeed, the type, quantity and quality of the exploited resources, both tangible and intangible, significantly affect the workability of the firm, since they place technical and organizational constraints. However, a key factor is the aptitude of the team to achieve cooperation and coordination in order to handle the flow of information. This kind of synergy is essential in the design practice, even more if considering sustainable design. Contrary to other industrial sectors where different issues are managed more or less independently by a specific team or experts, environmental issues involve all stakeholders of the building supply chain, with significant impact within the decision-making process. In this way, starting from the early stage of the project, collaboration, coordination and communication play a pivotal role in making sure that firm resources turn first into "capabilities", i.e. minimum ability, and later into "maturity", i.e. quality achieved by good practice.

The challenge is to orient and streamline the AEC design process in line with environmental targets and life cycle perspective. In fact, due to the high impacts generated on the environment and the strong international pressure of Sustainable Development Goals to be met by 2030, it is extremely compelling to appropriately change the attitude and way of practice of the building sector. To comply with the shared environmental targets, it is crucial to drive and orient in the right direction the construction sector, to achieve great results in a limited time span. The required environmental-oriented reshaping represents a great challenge for the construction sector, demanding an even more robust transformation process in a sector till now considered resistant to change. Moreover, the change management calls for the involvement of all the actors responsible for the built environment, starting from the mindset to end with practical actions and operations.

Consistent with the trends currently underway that lead to consider artefacts as a small part of a larger networks, systems and environment, two are the shared strategies that need to be applied. Firstly, the implementation of Life Cycle Thinking (LCT) to change the mindset of the construction sector practitioners. Secondly, the application of the Life Cycle Assessment (LCA) method to evaluate, adjust and verify the way of acting within design and construction practices, endorsing a holistic approach. For this purpose, the book intends to support AEC practice in the implementation of LCT and then, in the next future, of LCA, to allow practitioners to make aware decisions, to avoid shifting problems, to gain a long-term perspective, to optimize design process, to lead decision-making and to effectively decrease construction impacts.

In order to fill the gap between theory and practice, the methodology adopted combines the literature research, developed by exploring the studies and findings available in the scientific and public communities, with the field research, developed by conducting interviews with design and construction firms. In particular, the working method elected for closely investigating current AEC firms and their workability refers to the qualitative ethnographic approach. Indeed, ethnography is now emerging as part of the set of techniques used to understand the construction industry, a sector considered extremely complex and influential but that despite this remains mostly unexplored and under-theorised. In this way, the belief is that construction ethnography and surveying the main actors engaged in the process offer new routes to knowledge about and in the construction sector.

Ethnography represents thus the methodology adopted for the search on the field, applying two different models of interviews in relation to the level of detail to be achieved. The first model concerns the spread of a questionnaire survey, involving a large target audience, focusing on a set of design and construction practices and using indirect means of communication. By contrast, the second model implies a personal involvement within joined AEC practices, establishing specific partnerships, focusing on a sample of environment-friendly projects and using direct means of communication.

To figure out how AEC firms are equipping and reorganizing themselves in order to address and meet environmental and life cycle issues, the book moves from the general to the particular. It begins to question if environmental issues are really changing and shaping AEC firms, finding answers both at theoretical and practical levels. On one side, an extensive literature review is developed, providing an overview of the phenomena ongoing in the building sector (Chap. 1) and focusing thereafter on the key emerging trends to address environmental sustainability, including BIM, Green BIM and LCA (Chap. 2), the related tools and methods to support decision-making (Chap. 3) and the building-related LCA applications (Chap. 4). On the other side, the topic is explored in the field, interviewing design and construction companies through a questionnaire survey (first model of interview), in order to start gaining insight into the AEC perspective and providing a synopsis of the inherent environmental-driven change management (Chap. 5).

Once attested the reshaping of AEC firms towards environmental sustainability, the focus is on how environmental and life cycle issues affect the AEC decision-making process. Consistent with the trends currently underway in the building sector, this objective is combined with the challenge to implement LCT and related methodologies within the AEC practice as a turning point. For this reason, an assessment framework is presented to match life cycle perspective and the building process, focusing on the set of the required information, involved actors and used tools (Chap. 6). Based on this, a shared framing is developed for outlining the environmental and life cycle efforts faced in practice by design firms, identifying the resources invested along the design process (Chap. 7). By its application on a sample of environment-friendly projects selected from the portfolio of established A/E firms (second model of interview), the design process of current practice is mapped ex-post according to different viewpoints. In particular, the involved design team directory (Chap. 8), the performed environmental and life cycle studies (Chap. 9), the pursued Green Building Rating System certification process (Chap. 10) and the set of life cycle information considered during the decision-making (Chap. 11) are analysed in depth to provide hotspots towards life cycle design.

Finally, starting from the existing information flow implemented by design team to meet environmental and life cycle issues (Chap. 12), it is envisioned the change management for switching from current design practice to life cycle AE(C) practice, supposing to apply the proposed assessment framework starting from the early phase of the process. To facilitate a consistent implementation in practice, personal worksheets are inferred from the framework for entrusting the involved competences in designing the committed technological elements according to life cycle criteria, progressively collecting the relative information (Chap. 13). In this way, the information flow demanded for life cycle-oriented decision-making process, conceived in a BIM-oriented working environment, results in a project-based and well-framed set of data of the facility during the whole life cycle, representing a significant added value and a continuous build-up of know-how (Chap. 14). To conclude, key findings, possible stakeholders and outlook for life cycle practice are detected (Chap. 15).

Milan, Italy Anna Dalla Valle

Contents

Chapter 1
Reshaping of AEC Firm Management to Face Environmental Sustainability

Abstract The current globalization of market, specialization of competencies, digitalization of the process and wide range of requirements make construction sector increasingly complex, gradually changing Architecture, Engineering and Construction (AEC) firms towards sustainability. To this end, an overview of the phenomena ongoing in the construction sector is provided, firstly, focusing on the overall change management of multidisciplinary firms and, secondly, on the leading environmental issues and tools implemented during the process. The transformation process involves both tangible and intangible resources and it is now supported by a broad spectrum of tools that assist practitioners in making better business decisions, especially towards environmental goals. Here, life cycle approaches are growing in importance to implement the whole-system thinking required in the decision-making process, to meet clients demands for environmentally friendly buildings as well as to develop environmental processes and practice, enhancing competitiveness, sustainability and innovation on the entire supply chain.

1.1 Complexity and Change Management of the Building Sector

Over the recent decades, building sector has expanded and become increasingly complex, as direct consequence of the globalization of market (Bond and O'Byrne 2014) and the demand of a wide range of requirements (Gong and Janssen 2013), such as the ones related to sustainable development currently spread worldwide. Faced with these pressures, the construction field is changing step by step (Renz and Solas 2016; Deamer and Bernstein 2010), even if AEC industry is still now considered resistant to change (Smollan 2011).

The transformation occurs both in design process and building construction, boosting the change management of AEC firms (Chinowsky and Byrd 2001), especially reshaping the role of architects (Manzoni et al. 2014; Moser 2014) and developing new collaboration models and delivery methods to create competitive advantage (Srivastava et al. 2013). In this way, the process of change involves all AEC firm assets: tangible resources, namely the physical items visible to everyone, such

A. Dalla Valle, *Change Management Towards Life Cycle AE(C) Practice*,
PoliMI SpringerBriefs, https://doi.org/10.1007/978-3-030-69981-9_1

as materials, buildings, plant, equipment, tools and money; and intangible resources, intending the nonphysical and invisible entities, such as knowledge, organization and competences of practitioners (Norsa 2005; Sinopoli 1997).

Concerning tangible resources, technology is always more relevant in AEC practice: the set of tools available on the market is able to meet any design issues, allowing new capabilities with significant impact on practices (Becerik-Gerber and Kensek 2010; Riese 2012). AEC industry appears thus even more digitalized and virtualized, even if technological transformations are difficult to be implemented in a successful way (Davis and Songer 2009). The so-called digital revolution started in the early 1990s exploiting digital technology for computerized technical image CAD 2D, then for virtual image CAD 3D and finally for information flow management BIM–2D-3D-4D- 5D-6D (Deamer and Bernstein 2010). The actual shifting from representation to simulation emphasises the transformative power of BIM (Gerbert et al. 2016) and the need to assist companies in its adoption and implementation (Khosrowshahi and Arayici 2012), identifying competencies and framework to organize domain knowledge (Succar 2009). In fact, integrating extensive building metadata and information (Boddy et al. 2007; Rezgui et al. 2011), BIM encompass the whole supply chain, changing the conventional design processes (Chi et al. 2014), linking production and construction (Babič et al. 2010) and generating optimized activity-level of construction schedules (Liu et al. 2015).

However, the current scenario pushes more and more toward the understanding that in a highly competitive context the real added value lies in intangible resources (Mills et al. 2003), dealing with knowledge and know-how. Indeed, to meet the wide range of requirements, on one hand, AEC firms require a growing division of labour and an increasingly specialization (Hoffman and Lintern 2006), on the other, they call for an integrated design process where individual efforts turn into systems to get sustainable and competitive advantage (Trebilcock 2009). To this end, centralized or decentralized hierarchies and internal or external equipment are adopted as appropriate by companies, strongly affecting the decision-making process (Wang 2011). Consequently, AEC firms invest in knowledge management throughout the project lifecycle (Kanapeckiene et al. 2010; Rezgui et al. 2010), significantly enhancing environmental practices (Reed et al. 2014), although demanding in some cases a knowledge intermediary to support function within value chain networks (Krenz et al. 2014).

All these pressures on both tangible and intangible resources determine within AEC firms the need of change management to improve productivity in the process and make better business decisions, promoting innovative and organizational changes (Lines et al. 2016; Weippert and Kajewski 2010). Indeed, companies have to plan for and react to change with the goal to deliver quality projects on-time and under budget, choosing between a wide variety of project delivery approaches to suit their needs, with statistically performance improvement in Integrated Project Delivery (El Asmar et al. 2013). For this purpose, a key factor is the ability of organisations to achieve collaboration, cooperation and coordination among the stakeholders of the supply chain, supported by various integration systems and collaboration technologies (Shen et al. 2010).

Knowledge and collaboration are therefore pivotal for success, resulting in projects life cycle management systems to integrate practitioners, data, processes and business as well as to share information and workflows within companies (Sakhare et al. 2014). Nevertheless, due to the complexity and fragmented nature of construction industry, the problem of data integration throughout the whole project life cycle still remains a challenge. Nowadays, it is used a cloud approach to link building data (Curry et al. 2013; Jiao et al. 2013), but also a cyber-physical systems approach to improve information exchange between virtual models and physical construction (Akanmu and Anumba 2015). In this context, interoperability represents a crucial task, on one side, due to the variety of applications and systems used by different players and, on the other, to the dynamics and flexibility needed to operate in building sector (Grilo and Jardim-Goncalves 2011; Tibaut et al. 2016).

1.2 Life Cycle Approach as Turning Point Towards Sustainability

The building sector is widely recognized as one of the most incisive and impacting sector at a global scale, covering all three pillars of sustainability. From the environmental point of view, it yearly consumes about 3 billion tons of raw materials and it is responsible for 40% of solid waste and for 25–40% of global energy use. From the economical point of view, it collects total annual revenues of $10 trillion, accounting for about 5% of GDP and employing more than 100 million people worldwide. From the social point of view, it steadily shapes our daily life with a high impact on health status and well-being of users (Renz and Solas 2016). In this context, it represents a strategic field of action in which promptly intervene for the achievement of Sustainable Development Goals within 2030 and the transition towards a circular economy.

Indeed, circular economy is gaining increasing attention at international level as a way to increase prosperity through new business models, while reducing dependence on primary materials and energy (Ellen MacArthur Foundation 2015). For this purpose, in building sector a whole-system and life cycle thinking is needed to reduce environmental impacts and drive both sustainability and innovation (Faludi 2015), starting at a large scale with supply chain to end at building level (Antink et al. 2014). Here, the implementation of life cycle approach represents a turning point not only to meet customer demands for environmentally friendly products and projects, but also to improve processes and services towards sustainability and thus increase competitiveness (Khasreen et al. 2009; Ortiz et al. 2009). In fact, Life Cycle Thinking (LCT) and related methodologies promote sustainable practice, encourage innovative strategies, lead decision-making, avoid shifting problems and support long-term perspectives.

In literature, many studies review LCA methodology in the construction sector, focusing on methods, applications and frameworks (Buyle et al. 2013), identifying the phases and materials that mostly affect the environment (Abd Rashid and Yusoff 2015). Going into detail, some works explore single LCA phase, for instance linking LCI and innovation (Luz et al. 2015) or comparing LCIA indicators and models (Frischknecht et al. 2016; Owsianiak et al. 2014). By contrast, other studies put the methods into practice, applying from-cradle-to-grave LCA analysis to building typologies (Asdrubali et al. 2013), to prefabricated building systems (Jaillon and Poon 2014) and to remediation projects (Favara et al. 2011), developing simplified or detailed LCA analyses (Kellenberger and Althaus 2009).

Besides LCA method, which evaluates all resources and flows, including energy, water and materials, and environmental loads, including CO_2 emissions, solid and liquid waste, other life cycle approaches are available for assessing the environmental impacts of building construction. To name a few, Life Cycle Energy Analysis (LCEA) is focused on energy input (Cabeza et al. 2014), Life Cycle Carbon Emissions Assessment ($LCCO_2A$) is focused on the CO_2 equivalent emissions (Baek et al. 2013), and Life Cycle Cost (LCC) is focused on the total cost of facility ownership (Chau et al. 2015). Furthermore, in order to promote a more holistic assessment of building life cycle performance, some methods combine LCA with LCC analysis (Gu et al. 2008), while others promote a whole life project management perspective to achieve the shared sustainable goals (Wang et al. 2014).

Here, building materials selection shows a growing importance not only on the performance of buildings but also on their impacts. To facilitate integrated design teams in the evaluation of building materials, life cycle-oriented criteria are developed to support the decision-making, matching for instance materials environmental performance with service life expectations (Haapio and Viitaniemi 2008; Mora et al. 2011) or with economic parameters (Takano et al. 2014).

In this context, clients undertake a crucial role in terms of creating and stimulating the right conditions for construction sustainability and innovation, understanding and sharing the needs of both end-users and stakeholders (Hartmann et al. 2006). On one side, they are in charge of the delivery process of design-build projects, by promoting new strategies, best practice and innovative management approaches (Kilinc et al. 2015). On the other, clients assimilate building regulations and governmental policies for minimizing construction environmental impacts, becoming one of the most important drivers for sustainable buildings (Häkkinen and Belloni 2011; Ozorhon 2013; Shaw and Ozaki 2016).

1.3 Sustainable Design Practices and Environmental Tools

Sustainable design is an expression commonly used in the realms of design practice but, since its definition remains quite fuzzy, it is important to clarify his meaning in the AEC domain (Sharma 2010). Nowadays, there are several barriers in sustainable building, such as the ability to deliver green project within acceptable cost constraints, where the key to success depends on the involvement of a cross-discipline design team starting from the early project phases (Robichaud and Anantatmula 2011) and on the establishment of non-linear and retroactive process (Mendler et al. 2006). However, driven by sustainable development, in the next future every design and construction firm need to transform itself to survive and succeed, aiming at integration, transparency, accessibility and collaborative learning (Kaatz et al. 2006).

To his end, several works available in literature present methods, frameworks and models to support and advance building practice in fostering sustainable constructions. For instance, a four-stage model is developed for companies to progress initially from engaging with sustainability to accelerating, leading and ultimately, transforming their businesses (Hedstrom 2015), choosing among different possible ways (Farmer 2013). Another study, based on the triple bottom line principle, allows design teams to obtain a proper balance between resource conservation, cost efficiency and design for human adaptation (Akadiri et al. 2012). Other frameworks are developed with an integrative system approach to enable reorganization and integration, emphasizing the perspective of decision-makers and stakeholders (Mukherjee and Muga 2010) and the awareness of project participants (Irizarry et al. 2012).

Furthermore, questionnaires are carried out over the years to survey sustainable practice in construction sector. They regard for instance the design process of advanced practice when executing sustainable design, focusing on Building Information Model (BIM), Integrated Project Delivery (IPD), Integrated Design Process (IDP) and Building Energy Simulation (BES) (Farias Stipo 2015). Other authors tried to understand the role played by the different professionals if the company decide to move toward sustainability (Nemerson and Rosenbeck 2012), as main responsible to sustainability efforts, generally intended as use of less energy and natural resources, reduction of emissions and material waste, and adoption of renewable, recyclable and recycled materials (Rosen 2013). Lastly, an interview investigates users' perceptions of sustainable buildings, stressing the highly commitment of clients to the sustainable practice, involved in green business and endowed with long-term perspectives (Baird 2015).

The challenge of sustainability is not addressed only by AEC firms but also by software corporations (Autodesk 2015), spreading worldwide the set of sustainability assessment tools (Ding 2008) for their ability to evaluate the environmental performance of products and buildings (Berardi 2012), facing complex issues while remaining simple and practical (Cole 2005). Since a broad range of instruments is now available on the market for environmental improvement in buildings, many studies clarified the field describing, categorising and comparing existing environmental tools (Forsberg and von Malmborg 2004; Haapio and Viitaniemi 2008) to provide data for decision support (Zhai and McNeill 2014). However, dealing with quantitative and qualitative data, it emerges the issue of subjectivity by different stakeholders and decision-makers (AlWaer et al. 2008). In environmental tools the set of sustainability indicators are typically developed by highly trained experts, joining only in few cases participatory processes (Fraser et al. 2006). Their application encompasses more or less every field assessing sustainability at broad level in supply chain context (Olson et al. 2011), at intermediate level depicting corporate performance (Rahdari and Anvary Rostamy 2015) and at confined level evaluating building projects (Butler et al. 2011).

In this context, the term Green BIM is socially spread in building and construction sector to indicate the use of BIM tools to achieve sustainability and improve building performance throughout the project life cycle (see Chap. 2). If, on one side, architects and engineers perceive great value in using BIM for green projects, on the other, they need to be trained in order to integrate it in the workflow (Levring and Nielsen 2011), starting from the early design phases. Indeed, it allow to perform complex building performance analyses (Liu et al. 2015), in particular daylight and energy analysis (Gourlis and Kovacic 2016; Moakher and Pimplikar 2012), although there is the need to fill the performance gap between design simulation and real consumption by closing the feedback loop (Austin 2013; Motawa and Carter 2013). Moreover, Green BIM is currently used to develop different rating systems (Biswas et al. 2013), especially focusing on LEED certification process (Azhar et al. 2011; Jalaei and Jrade 2015; Wu and Issa 2015). Finally, whereas construction sector increasingly emphasized sustainability and Life Cycle Assessment is emerging as one of the most functional aid to the decision-making process (Bayer et al. 2010), the challenge is to achieve LCA-BIM integration for providing designers immediate feedbacks considering the whole life cycle (see Chap. 3).

References

Abd Rashid AF, Yusoff S (2015) A review of life cycle assessment method for building industry. Renew Sustain Energy Rev 45:244–248

Akadiri PO, Chinyio EA, Olomolaiye PO (2012) Design of a sustainable building: a conceptual framework for implementing sustainability in the building sector. Buildings 2:126–152

Akanmu A, Anumba CJ (2015) Cyber-physical systems integration of building information models and the physical construction. Eng Constr Architect Manag 22(5):516–535

AlWaer H, Sibley M, Lewis J (2008) Different stakeholder perceptions of sustainability assessment. Archit Sci Rev 51(1):48–59

Antink R, Garrigan C, Bonetti M, Westaway R (2014) Greening the building supply chain. UNEP. https://www.unep.org/sbci/pdfs/greening_the_supply_chain_report.pdf

Asdrubali F, Baldassarri C, Fthenakis V (2013) Life cycle analysis in the construction sector: guiding the optimization of conventional Italian buildings. Energy Buildings 64:73–89

Austin B (2013) The performance gap: causes and solutions. https://www.greenconstructionboard.org/images/stories/pdfs/performance-gap/2013-03-04ClosingtheGap_FinalReport_ISSUE.pdf

Autodesk (2015) Sustainability in action: from epic challenges to integrated solutions. https://damassets.autodesk.net/content/dam/autodesk/www/sustainability/docs/pdf/sustainability_report_2015_FINAL.pdf

Azhar S, Carlton WA, Olsen D, Ahmad I (2011) Building information modelling for sustainable design and LEED rating analysis. Autom Constr 20:217–224

Babic NC, Podbreznik P, Rebolj D (2010) Integrating resource production and construction using BIM. Autom Constr 19:539–543

Baek C, Park SH, Suzuki M, Lee SH (2013) Life cycle carbon dioxide assessment tool for buildings in the schematic design phase. Energy Build 61:275–287

Baird G (2015) Users' perceptions of sustainable buildings: key findings of recent studies. Renew Energy 73:77–83

Bayer C, Gamble M, Gentry R, Joshi S (2010) AIA Guide to building Life Cycle Assessment in practice. https://www.aia.org/aiaucmp/groups/aia/documents/pdf/aiab082942.pdf

Becerik-Gerber B, Kensek K (2010) Building information modelling in architecture, engineering, and construction: emerging research directions and trends. J Profess Issues Eng Educ Pract 136(3):139–147
Berardi U (2012) Sustainability assessment in the construction sector: rating systems and rated buildings. Sustain Dev 20:411–424
Biswas T, Wang TH, Krishnamurti R (2013) From design to pre-certification using Building Information Modelling. J Green Build 8(1):151–176
Boddy S, Rezgui Y, Cooper G, Wetherill M (2007) Computer integrated construction: a review and proposals for future direction. Adv Eng Softw 38:677–687
Bond C, O'Byrne DJ (2014) Challenges and conceptions of globalization: an investigation into models of global change and their relationship with business practice. Cross Cultural Manag 21(1):23–38
Butler PB, Laresen-Hallock L, Lewis P, Glenn C, Armstead R (2011) Metrics for integrating sustainability evaluations into remediation projects. Remediation J, 81–87
Buyle M, Braet J, Audenaert A (2013) Life cycle assessment in the construction sector: a review. Renew Sustain Energy Rev 26:379–388
Cabeza LF, Rincon L, Vilarino V, Pérez G, Castell A (2014) Life cycle assessment (LCA) and life cycle energy analysis (LCEA) of buildings and the building sector: a review. Renew Sustain Energy Rev 29:394–416
Chau CK, Leung TM, Ng WY (2015) A review on life cycle assessment, life cycle energy assessment and life cycle carbon emissions assessment on buildings. Appl Energy 143:395–413
Chi HL, Wang X, Jiao Y (2014) BIM-enabled structural design: Impacts and future developments in structural modelling, analysis and optimisation processes. Arch Comput Methods Eng 22:135–151
Chinowsky PS, Byrd MA (2001) Strategic management in design firms. J Profess Issues Eng Educ Pract 127(1):32–40
Cole RJ (2005) Building environmental assessment methods: redefining intentions and roles. Build Res Inform 33(5):455–467
Curry E, O'Donnell J, Corry E, Hasan S, Keane M, O'Riain S (2013) Linking building data in the cloud: integrating cross-domain building data using linked data. Adv Eng Inform 27:206–219
Davis KA, Songer AD (2009) Resistance to IT change in the AEC industry: are the stereotypes true? J Constr Eng Manag 135(12):1324–1333
Deamer P, Bernstein P (2010) Building (in) the future: recasting labour in architecture. Priceton Architectural Press, New York
Ding GKC (2008) Sustainable construction-the role of environmental assessment tools. J Environ Manage 86:451–464
El Asmar M, Hanna AS, Loh W (2013) quantifying performance for the integrated project delivery system as compared to established delivery systems. J Constr Eng Manag 139(11):1–14
Ellen MacArthur Foundation (2015) Towards a Circular Economy: business rationale for an accelerated transition. https://www.ellenmacarthurfoundation.org/assets/downloads/TCE_Ellen-MacArthur-Foundation_9-Dec-2015.pdf
Faludi J (2015) A sustainable design method acting as an innovation tool. Smart Innov Syst Technol 35:201–212
Farias Stipo FJ (2015) A standard design process for sustainable design. Procedia Comput Sci 52:746–753
Farmer G (2013) Re-contextualising design: three ways of practising sustainable architecture. Archit Res Quart 17(2):106–119
Favara PJ, Krieger TM, Boughton B, Fisher AS, Bhargava M (2011) Guidance for performing footprint analyses and life-cycle assessments for the remediation industry. Remediation J, 39–79
Forsberg A, von Malmborg F (2004) Tools for environmental assessment of the built environment. Build Environ 39:223–228

Fraser EDG, Dougill AJ, Mabee WE, Reed M, McAlpine P (2006) Bottom up and top down: analysis of participatory processes for sustainability indicator identification as a pathway to community empowerment and sustainable environmental management. J Environ Manage 78:114–127

Frischknecht R, Fantke P, Tschümperlin L, Niero M, Antón A, Bare J et al (2016) Global guidance on environmental life cycle impact assessment indicators: progress and case study. Int Journal of Life Cycle Assess 21:429–442

Gerbert P, Castagnino S, Rothballer C, Renz A, Filitz R (2016) Digital in engineering and construction: The transformative power of Building Information Modelling. BGC. https://futureofconstruction.org/content/uploads/2016/09/BCG-Digital-in-Engineering-and-Construction-Mar-2016.pdf

Gong Y, Janssen M (2013) An interoperable architecture and principles for implementing strategy and policy in operational processes. Comput Ind 64:912–924

Gourlis G, Kovacic I (2016) Building Information Modelling for analysis of energy efficient industrial buildings: a case study. Renewable and Sustain Energy Rev, 1–11

Grilo A, Jardim-Goncalves R (2011) Challenging electronic procurement in the AEC sector: a BIM-based integrated perspective. Autom Constru 20:107–114

Gu L, Lin B, Zhu Y, Gu D, Huang M, Gai J (2008) Integrated assessment method for building life cycle environmental and economic performance. Build Simul 1:169–177

Haapio A, Viitaniemi P (2008) A critical review of building environmental assessment tools. Environ Impact Assess Rev 28:469–482

Haapio A, Viitaniemi P (2008) Environmental effect of structural solutions and building materials to a building. Environ Impact Assess Rev 28:587–600

Häkkinen T, Belloni K (2011) Barriers and drivers for sustainable building. Build Res Inform 39(3):239–255

Hartmann A, Dewulf G, Reymen I (2006) Understanding the innovation adoption process of construction clients. Industry and technology diffusion. https://www.academia.edu/698865

Hedstrom GS (2015) Navigating the sustainability transformation. In: Conference T (ed) Notes D. Board, Trusted Insights for Business Worldwide, pp 1–16

Hoffman R, Lintern G (2006) Eliciting and representing the knowledge of experts. In: Ericsson KA, Charness N, Feltovich P, Hoffman R (eds) Cambridge handbook of expertise and expert performance. Cambridge University Press, New York, pp 203–222

Irizarry J, Zolfagharian S, Nourbakhsh M, Zin RM, Jusoff K, Zakaria R (2012) The development of a sustainable-construction planning system. J Inform Technol Constr 17:162–178

Jaillon L, Poon CS (2014) Life cycle design and prefabrication in buildings: a review and case studies in Hong Kong. Autom Constr 39:195–202

Jalaei F, Jrade A (2015) Integrating Building Information Modelling (BIM) and LEED system at the conceptual design stage of sustainable buildings. Sustain Cities Soc 18:95–107

Jiao Y, Wang Y, Zhang S, Li Y, Yang B, Yuan L (2013) A cloud approach to unified lifecycle data management in architecture, engineering, construction and facilities management: integrating BIMs and SNS. Adv Eng Inform 27:173–188

Kaatz E, Root DS, Bowen PA, Hill RC (2006) Advancing key outcomes of sustainability building assessment. Build Res Inform 34(4):308–320

Kanapeckiene L, Kaklauskas A, Zavadskas EK, Seniut M (2010) Integrated knowledge management model and system for construction projects. Eng Appl Artif Intell 23:1200–1215

Kellenberger D, Althaus HJ (2009) Relevance of simplifications in LCA of building components. Build Environ 44:818–825

Khasreen MM, Banfill PFG, Menzies GF (2009) Life-cycle assessment and the environmental impact of buildings: a review. Sustainability 1:674–701

Khosrowshahi F, Arayici Y (2012) Roadmap for implementation of BIM in the UK construction industry. Eng Constr Archit Manag 19(6):610–635

Kilinc N, Basak G, Yitmen I (2015) The changing role of the client in driving innovation for design-build projects: stakeholders' perspective. Procedia Econ Finance 21:279–287

Krenz P, Basmer S, Buxbaum-Conradi S, Redlich T, Wulfsberg JP (2014) Knowledge management in value creation networks: establishing a new business model through the role of a knowledge-intermediary. Procedia CIRP 16:38–43

Levring A, Nielsen D (2011) Schematic strategies and workflows for sustainable design development. https://it.scribd.com/document/246973469

Lines BC, Sullivan KT, Wiezel A (2016) Support for organizational change: change-readiness outcomes among AEC project teams. J Constr Eng Manag 142(2):1–11

Liu H, Al-Hussein M, Lu M (2015) BIM-based integrated approach for detailed construction scheduling under resource constraints. Autom Constr 53:29–43

Liu S, Meng X, Tam C (2015) Building information modelling based building design optimization for sustainability. Energy Build 105:139–153

Luz LM, De Francisco AC, Piekarski CM (2015) Proposed model for assessing the contribution of the indicators obtained from the analysis of life-cycle inventory to the generation of industry innovation. J Cleaner Production 96:339–348

Manzoni B, Caporarello L, Saviozzi FA (2014) L'architetto: sette sfide manageriali per la crescita professionale. Milano, Egea

Mendler S, Odell W, Lazarus MA (2006) The HOK guidebook to sustainable design. Wiley, New Jersey

Mills J, Platts K, Bourne M (2003) Competence and resource architectures. Int J Oper Production Manag 23(9):977–994

Moakher PE, Pimplikar SS (2012) Building Information modelling (BIM) and sustainability: using design technology in energy efficient modelling. J Mech Civil Eng 1(2):10–21

Mora R, Bitsuamlak G, Horvat M (2011) Integrated life-cycle design of building enclosures. Build Environ 46:1469–1479

Moser C (2014) Architecture 3.0: The disruptive design practice handbook. Routledge: New York

Motawa I, Carter K (2013) Sustainable BIM-based evaluation of buildings. Procedia Soc Behav Sci 74:419–428

Mukherjee A, Muga H (2010) An integrative framework for studying sustainable practices and its adoption in the AEC industry: a case study. J Eng Tech Manage 27:197–214

Nemerson D, Rosenbeck J (2012) A designer's role in sustainability: The extent to which product and process designers can play a role in environmental sustainability and social responsibility. https://www.idsa.org/sites/default/files/Nemerson.pdf

Norsa A (2005) La gestione del costruire: tra progetto, processo e contratto. Milano, Franco Angeli

Olson EC, Haapala KR, Okudan GE (2011) Integration of sustainability issues during early design stages in a global supply chain context. Artif Intell Sustain Design Symp 84–90

Ortiz O, Castells F, Sonnemann G (2009) Sustainability in the construction industry: a review of recent developments based on LCA. Constr Build Mater 23:28–39

Owsianiak M, Laurent A, Bjørn A, Hauschild MZ (2014) IMPACT 2002+, ReCiPe 2008 and ILCD's recommended practice for characterization modelling in life cycle impact assessment: A case study-based comparison. Int J Life Cycle Assess 19:1007–1021

Ozorhon B (2013) Response of construction clients to low-carbon building regulations. J Constr Eng Manag 139(12):1–10

Rahdari AH, Anvary Rostamy AA (2015) Designing a general set of sustainability indicators at the corporate level. J Cleaner Product 108:757–771

Reed MS, Stringer LC, Fazey I, Evely AC, Kruijsen JHJ (2014) Five principles for the practice of knowledge exchange in environmental management. J Environ Manage 146:337–345

Renz A, Solas MZ (2016) Shaping the future of construction: a breakthrough in mindset and technology. World Economic Forum. https://www3.weforum.org/docs/WEF_Shaping_the_Future_of_Construction_full_report_.pdf

Rezgui Y, Boddy S, Wetherill M, Cooper G (2011) Past, present and future of information and knowledge sharing in the construction industry: towards semantic service-based e-construction? CAD Comput Aided Design 43:502–515

Rezgui Y, Hopfe CJ, Vorakulpipat C (2010) Generations of knowledge management in the architecture, engineering and construction industry: An evolutionary perspective. Adv Eng Inform 24:219–228
Riese M (2012) Technology-augmented changes in the design and delivery of the built environment. Commun Comput Inform Sci 242:49–69
Robichaud LB, Anantatmula VS (2011) Greening project management practices for sustainable construction. J Manage Eng 27(1):48–57
Rosen MA (2013) Engineering and sustainability: attitudes and actions. Sustainability 5:372–386
Sakhare KN, Dabade BM, Kadu A (2014) PLM: Change management process with custom workflow template. Int J Innov Res Sci Eng Technol 3(3):2677–2680
Sharma A (2010) Contemporary practice in sustainable design: appraisal and articulation of emerging trend. WIT Trans Ecol Environ 128:119–129
Shaw I, Ozaki R (2016) Emergent practices of an environmental standard. Sci Technol Human Values 41(2):219–242
Shen W, Hao Q, Mak H, Neelamkavil J, Xie H, Dickinson J et al (2010) Systems integration and collaboration in architecture, engineering, construction, and facilities management: a review. Adv Eng Inform 24:196–207
Sinopoli N (1997) La tecnologia invisibile. Milano, Franco Angeli Editore
Smollan RK (2011) The multi-dimensional nature of resistance to change. J Manag Organ 17:828–849
Srivastava M, Franklin A, Martinette L (2013) Building a sustainable competitive advantage. J Technol Manag Innov 8(2):47–60
Succar B (2009) Building information modelling framework: a research and delivery foundation for industry stakeholders. Autom Constr 18:357–375
Takano A, Winter S, Hughes M, Linkosalmi L (2014) Comparison of life cycle assessment databases: a case study on building assessment. Build Environ 79:20–30
Tibaut A, Rebolj D, Perc MN (2016) Interoperability requirements for automated manufacturing systems in construction. J Intell Manuf 27:251–262
Trebilcock M (2009) Integrated design process: new paradigms in sustainable architecture. Arquitetura Revista 5(2):65–75
Wang N, Wei K, Sun H (2014) Whole life project management approach to sustainability. J Manag Eng 30(2):246–255
Wang Q (2011) A scenario simulation study of decentralization on Architecture, Engineering and Construction. In Engineering project organizations conference: Colorado, 1–21
Weippert A, Kajewski S (2010) Managing innovative change within organisations and project team environments. Challenges, opportunities and solutions in structural engineering and construction, 823–828
Wu W, Issa RRA (2015) BIM execution planning in green building projects: LEED as a use case. J Manag Eng 31(1):1–18
Zhai ZJ, McNeill JS (2014) Roles of building simulation tools in sustainable building design. Build Simul 7:107–109

Chapter 2
Emerging Trends and Developments in BIM, Green BIM and LCA

Abstract In the light of AEC firms change management, the key emerging trends are surveyed fostering practitioners point of view. In particular, the focus is on the current state of Building Information Modelling (BIM), Green BIM and Life Cycle Assessment (LCA) in practice. Indeed, BIM is rapidly transforming the global construction industry representing the starting point for the digital transformation, being spread worldwide whilst with different level of maturity. Moreover, to meet sustainability, BIM is increasingly used to improve building performance and to encourage integrated design and construction processes over the building life cycle, turning out in Green BIM. Finally, practitioners start experiencing LCA in practice, since internationally recognized as the most scientific methodology to assess the environmental impacts of products, components and buildings throughout the whole life cycle.

2.1 BIM Business Value, Trends in Action, Implementation

As proved, nowadays Architecture, Engineering and Construction (AEC) firms are facing enormous technological and organizational changes and challenges, due to the proliferation of information technology in conjunction with the ambition to applicate sustainable practice (see Chap. 1). In this context, Building Information Modelling (BIM) is rapidly transforming the global construction industry, being increasingly spread worldwide, even if with different level of maturity, representing the starting point for an integrated digital transition, that is revolutionizing almost all aspects of building sector.

To deepen the topic from AEC practitioners' point of view, the questionnaire surveys conducted in literature are collected and analysed in-depth to provide an overview of the current state of BIM in practice (Becerik-Gerber and Kensek 2010; Khosrowshahi and Arayici 2012; McGraw-Hill Construction 2009, 2010). In fact, the reference surveys involve a wide range of professionals, including mainly architects, engineers and contractors, followed by other building stakeholders, such as owners, planners, construction product manufactures, government agencies, various integrated firms and consultants, exploring in particular USA and Europe contexts. The analysis shows that, although the sourced questionnaires are distinguished by very

A. Dalla Valle, *Change Management Towards Life Cycle AE(C) Practice*,
PoliMI SpringerBriefs, https://doi.org/10.1007/978-3-030-69981-9_2

different rate of response, respectively 54 (Becerik-Gerber and Karen Kensek 2010), 32 (Khosrowshahi and Arayici 2012), 2.228 (McGraw-Hill Construction 2009) and 948 responses (McGraw-Hill Construction 2010), they agree on how BIM is esteemed and exploited by AEC firms.

Results prove that Autodesk Revit turns out to be the most popular BIM tool, followed by ArchiCAD and Bentley and that, among building practitioners, architects have the highest level of BIM adoption and proficiency, followed in turn by engineers and to a lesser extent by contractors. BIM users declare to gather benefits throughout the project life cycle, whilst experiencing greater value in some phases rather than others. In particular, they benefit more during the design development phase and construction document phase, as BIM capabilities become notable in conjunction with more detailed models. Moreover, proceeding from the highest to the lowest perceived value, BIM users take high advantages during the concept phase, where BIM allows the design team to jointly work with clients and stakeholders to define project parameters. For parts of users, BIM provides significant value also during the production phase and bidding phase, since it helps in managing all products information and data, including specifications for contractors, fostering communication between design and building team. In addition, BIM enhances the construction phase because, reducing the number of building elements clashes, it helps to better control project budget and schedule, saving time and money. Finally, data-rich BIM models have positive impact on manufacturing phase, since producers can directly extract data rather than draw specifications into their existing systems. However, BIM value depends strictly on the types of users. Architects identify the highest value during the design development and construction document phases, but also during the concept. By contrast, engineers and contractors experience great value in tender and construction phases.

Furthermore, the questionnaire surveys display the multiple benefits of BIM for building design team, increasing its rate of success within collaborative working environments. Indeed, to maximise BIM opportunities, practitioners often implement collaborative and integrated approach to design and build, rethinking traditional roles and workflows to enable efficient communication and information exchange, reducing errors and improving productivity. Based on replies, BIM provides a broad spectrum of business benefits for all parties involved in the entire building supply chain. All respondents indicated that the most important advantage is the improvement of the collective understanding of design intent, supported by 3D visualizations and the rich database of project information. Indeed, BIM allows the project team to perform quick analysis of alternatives, to better interpret and verify project proposal and manage client expectations. Moreover, it assists in improve communication among all design and construction parties, by constantly sharing data with the allowed practitioners and avoiding data re-entering. BIM virtual design and construction enable also to identify problems earlier in the building process, reducing rework and errors and, by exchanging interoperable models and data, ensuring complete and accurate information for the construction documents. The improved accuracy of documents leads most of BIM users to detect that the cumulative BIM benefits across project life cycle are noted by owner in the building outcome and in the overall project quality. In particular, contractors and engineers recognize that the reduction of conflicts and changes during construction has great rewards on the project, affecting both building budget and schedule. Moreover, BIM allows to considerably reduce project delivery times between the design and construction phases, to increase the prefabrication of components and to support the building asset management along the whole life cycle. Furthermore, many BIM users stress their capability to capture building projects that require BIM competences as well as the chance to introduce BIM technology to new clients, gaining competitive advantage.

Concerning non-users, the conducted questionnaires disclose that the primarily reason for not implementing BIM is the lack of clients demand, followed by the lack of time to evaluate it and by the high price of software. However, architects, engineers and contractors agree on the increasingly importance of BIM over the coming years, suggesting some areas of improvement. All users at all experience levels recognize the urgent need for improving interoperability, since its limits BIM potentialities and data exchange among tools, although efforts to create standards are underway. In addition, they solicit the development of 3D products objects with embedded manufacturer content, in order to integrate them in the building model and improve conveyance of design ideas, more accurate energy simulations and earlier cost estimations. Moreover, other field of interest affirmed by practitioners are the enhancement of integrated project delivery and the exploitation of BIM for sustainable design and construction and for project life cycle. In particular, it emerges the need to maintain a single project information database from design to fabrication, installation, operations and maintenance. For this purpose, further advancements need to be accomplished to fully integrate a wide range of graphic and non-graphic data within building-specific database over the whole life cycle.

2.2 Green BIM Sustainable Design Practice and Potentials

Nowadays building sector is distinguished by two key trends that are rapidly transforming design and construction practices: on one hand, the implementation and spread of BIM and, on the other, the growing demand for sustainable buildings. As logical development, project teams are incrementally exploring how BIM helps and supports the achievement of more sustainable outcomes. Indeed, BIM is able to provide promptly insight into how design decisions impact building performance and to encourage integrated design process, in either case critical strategies towards environmentally sustainability. In this way, it emerges a correlation between the establishment of sustainable practice and the adoption of BIM, because of BIM facilitates green design and construction and, simultaneously, the growth of sustainable buildings helps in accelerating BIM application.

Due to lack of commonly accepted definition, authors define Green BIM in different ways, stressing some aspects and potentialities rather than others. In broad terms, Green BIM is considered the use of BIM tools to achieve sustainability and improve building performance of projects (McGraw-Hill Construction 2010) or, similarly, as the synergies of BIM and green building to improve sustainable outcomes (Wu and Issa 2015). In addition, some authors underline that Green BIM support building design industry to efficiently integrate sustainable components into the building project lifecycle (Alawini et al. 2013). Others define it as a model-based process of generating and managing coordinated and consistent building data during its project lifecycle, enhancing building energy-efficiency performance and facilitating the compliance with sustainability goals (Wong and Zhou 2015).

To figure out the current state and benefits of Green BIM, practitioners' point of view is explored through the analysis of the questionnaire surveys available in literature (Azhar and Brown 2009; Bynum et al. 2013; McGraw Hill Construction 2010). The collected surveys are characterized by a shared target audience which includes: architectural firms, construction companies, engineering firms and other industry respondents, like owners, building product manufacturers, government agencies, integrated firms and consultants. By contrast, they are distinguished by different response rate, gathering as appropriate 91 (Azhar and Brown 2009), 123 (Bynum et al. 2013) and 494 responses (McGraw-Hill Construction 2010), referred especially to USA context. Matching the set of questions and answers, an overview of the perception of Green BIM in practice is offered, exploring BIM-based sustainability analyses as well as the expected future applications.

Practitioners are aware of the capabilities of BIM to enable highly sustainable outcomes. As results, Green BIM professionals use the tools more often, more fully and on more projects than non-Green BIM companies, showing how there is greater BIM experience and sophistication among firms who practice Green BIM. The key BIM features ranked by users in terms of sustainability are: the direct integration with energy analysis software, project management software and construction-related tools, such as cost estimation, scheduling and quantity take-off; the direct creating of construction documents; the ability to support multidisciplinary, collaborative and integrated workflow and to facilitate interoperability. Moreover, all respondents agree that the two top drivers to start in Green BIM practice are owner demand and time and cost savings, followed by the design firm willingness to promote the use of BIM on sustainable building projects. Conversely, most of the firms not practicing Green BIM declare that they do not employ it for a variety of reasons, including lack of client demand and fewer green projects in their work.

Green BIM practitioners usually dealt in practice with an increased number of green projects compared to non-Green BIM companies, showing great experience with green work. Moreover, since the majority of interviewed firms claim to be specialized in green buildings, the surveys reveal their penetration into the market, with a high involvement especially of A/E firms rather than contractors, even if their engagement is growing over time. Nearly all firms involved in Green BIM use BIM for new design and construction projects, followed by major retrofit projects and to a lesser extent by small retrofit projects and maintenance activities, leading to consider its suitability to larger and more complex projects. In addition, the small share of maintenance operations is not surprising, because most respondents are design firms and contractors, not in charge of service activities. Concerning the project phases in which Green BIM is typically performed, most users indicated that BIM-based sustainability analyses are carried out mainly during the design process, followed in turn by concept and construction phases and marginally by post-construction. Anyway, all respondents jointly assert that the optimal phases for implementing BIM to ensure sustainable practice is the schematic design phase, stressing therefore the adoption starting from the early phases of the process.

Deepening the performed sustainability analyses, Green BIM practitioners declare that the most common studies regard energy consumption, including energy simulations, lighting and HVAC analyses, followed by building orientation, massing and solar analyses. Moreover, the majority of respondents exploits BIM for setting up green building rating system documents, building product materials evaluations and, especially for contractors, bill of quantities. On the contrary, the analyses less affirmed within a Green BIM working environment turn out to be: renewable energy, carbon emission, acoustic and water use analyses. To perform the wide range of sustainability analyses, the tools commonly used in practice are, in order from most to less widespread: Autodesk Green Building Studio, IES Virtual Environment, Autodesk Ecotect, followed by Energy Plus, Delight, Radiance, HEED, Homer, Virtual DOE, Bentley Hevacomp and Climate Consultant.

In overall terms, industry practitioners agree that they are just beginning to explore full capabilities of Green BIM, demonstrating how they perceive great potential for BIM to improve design and construction practice as well as sustainable buildings. In particular, with the growing establishment of green building rating systems, such as LEED certification, Green BIM practitioners state that actually BIM has low level of automation to calculate credits. They demand hence for a broad spectrum of functionalities to help in verifying building performance and documenting LEED credits documentation more effectively, creating new opportunities for technology companies. Furthermore, all BIM practitioners consider green products as a challenging task for sustainable projects and as a topic of growing interest in practice, whether the firms use BIM for green projects or not. To this end, especially Green BIM users are aware of Life Cycle Assessment to measure products sustainability and they know the possible link between LCA and BIM. Despite they rarely take advantage of it, practitioners claim the importance of performing LCA-BIM, recommending the growth of BIM capabilities in this area.

2.3 LCA Current Status and Future Challenges

As design and construction industries increasingly emphasize sustainability, many methods were developed to evaluate and reduce the environmental impacts of buildings. In this context, Life Cycle Assessment (LCA) turns out to be the most scientific and recognized methodology established at international level to assess the environmental impacts of products, components and buildings throughout the whole life cycle. For this reason, its application in practice is even more considered as a turning point to promote sustainable and environmentally-friendly projects and it is therefore supported by several policies, norms and Green Building Rating Systems (EC-JRC

2017). However, if on one side it is clear the overall awareness of the issue within the scientific and public communities, on the other, there is little evidence of how LCA is perceived and experienced by building practitioners.

To understand the role of LCA in building sector and the related challenges and needs from the practitioners' point of view, the surveys available in literature are taken as a reference for the exploration (Han and Srebric 2015; Schlanbusch et al. 2016). The few gathered questionnaires refer to different context, target audience and response rate, but if combined they provide an overview of the current state of LCA in building practice. In particular, on one side, the survey was completed by 96 practicing designers in the US, specialized in enclosure systems and mechanical systems (Han and Srebric 2015). On the other, it interviewed 57 different stakeholders from the Nordic European Country with varying LCA expertise, including architects but also environmental and energy consultants, building product manufacturers, entrepreneurs, building owners, organizations and research institutes (Schlanbusch et al. 2016). Note that in both cases, building designers pertain to small- and medium-size firms, not involving big-size firms.

Up to now, LCA is still not widely used in building sector, being applied in a small percentage of design projects, especially in comparison with the now spread building energy simulations. As happens for both BIM and Green BIM, the main driver for performing LCA during design is the owner, stressing the importance of increasing clients' education on the benefits offered to achieve sustainable buildings. Other drivers are, in decreasing order of perceived impact: designers, codes, LEED and envelope consultants. Moreover, among components and materials selection criteria, environmental impacts are ranked with the lowest importance level as compared with other criteria, such as initial cost, energy efficiency, Life Cycle Costing, durability and maintenance.

Regarding methodology, LCA practitioners state that the main knowledge gaps in building sector are (from high to low): the environmental benefits of recycling, reuse and energy recovery; the lack of estimated default values on buildings parts or components for use in screening LCA; the environmental impact of maintenance and repair; the scenarios and building context; the environmental impact of demolition and reconstruction; the emission factors for energy sources and electricity; other impact categories than global warming; and lack of data for certain product categories. All these factors are considered by respondents as important challenges for performing LCA in buildings, prioritizing the environmental impacts associated to benefits and loads beyond the systems boundary. Concerning screening LCA, since the aim is to apply it starting from the early design phases when life cycle inventory and detailed products are not yet defined, users emphasize the need of default and reference values at least for the most impactful elements. The importance of LCA reference values is also mentioned for different types of buildings. In addition, some practitioners criticize the current available databases since incomplete and not updated, specify the relevance of other impact categories than global warming potential and remark the inclusion of harmful substances and chemicals in LCA studies.

To develop LCA study at building level, LCA practitioners ascertain missing data and assessment guidelines primarily for the use stage, including maintenance, repair, replacement and refurbishment processes (B2-B5) and to follow for the end of life stage (C1–C4), the benefits beyond the system boundary (D), the construction and installation stage (A5), the transportation to building site (A4) and lastly the product stage (A1-A3). In this way, the assumed scenarios are generally distinguished by missing data and guidelines, representing the major challenge in building LCA. In particular, difficulties are faced in dealing with service life and replacement cycles, claiming for a better understanding of LCA application on renovation projects and in the choice between deep renovation and new construction. Practitioners also stress that most EPDs declare environmental profile of products only referred to production stage (from cradle to gate), lacking assessment of the following stages, especially calling for the impacts of end of life stage and benefits and loads beyond the system boundaries.

As disclosed, the primary barrier faced by building practitioners in the application of LCA is the effort to find and collect data. Moreover, they indicate that performing a building complete LCA study during the decision-making process is time consuming as well as expensive, covering a very little market demand. Nonetheless, to encourage its implementation, practitioners recommend the development of simplified LCA tools in order to reduce time, especially for the inventory phase, and to use LCA during the design processes in a flexible way to really supports professionals in finding optimal solutions. With regard to LCA results, practitioners experience difficulties in comparing different products and context and in assessing the quality and uncertainty of the evaluation, due to lack of data transparency, shared guidelines and knowledge gaps, for instance on the impact that methodological choices have on results. Moreover, it emerges the need of digitalizing EPD data, now based on static pdf documents, in order to facilitate through interoperability their import in LCA tools and applications.

Based on these challenges, LCA practitioners solicit the collaborations among the interested stakeholders at national and international level. In particular, the creation of a common database appears of interest, to provide both reliable data and comparable LCA studies. In addition, priorities by decreasing order of importance are assigned to: the creation of shared criteria and requirements, the harmonization of existing tools and labels; and lastly the development of new tools. In this context, BIM is considered a promising tool for performing LCA studies of buildings, pointing at the integration between BIM and LCA to provide immediate feedback during the decision-making process. To this end, designers give prominence to the current lack of BIM in supporting the performance management of buildings as well as to the missing availability of product information and smart methods for data collection. Finally, they stress the chance to embed EPD data within BIM to encourage environmentally-friendly decisions starting from the early design phases and the development of guidelines for easily extracting and systematising data from BIM in order to save time and improve the quality of buildings LCA inventories.

References

Alawini A, Tanatanmatorn N, Tucker D, Tucker K, Daim T (2013) Technology adoption: building IT. In: Daim T, Oliver T, Kim J (Eds.). Research and technology management in the electricity industry. Springer: Switzerland

Azhar S, Brown J (2009) BIM for sustainability analyses. Inte J Constr Educ Res 5:276–292

Becerik-Gerber B, Kensek K (2010) Building information modelling in architecture, engineering, and construction: emerging research directions and trends. J Profess Issues Eng Educ Pract and Practice 136(3):139–147

Bynum P, Issa RRA, Olbina S (2013) Building information modelling in support of sustainable design and construction. J Constr Eng Manag 139(1):24–34

European Commission—Joint Research Center (2017) Level(s)–A common EU framework of core sustainability indicators for office and residential buildings. Publications Office of the European Union, Luxembourg

Han G, Srebric J (2015) Comparison of survey and numerical sensitivity analysis results to assess the role of life cycle analyses from building designers' perspectives. Energy Buildi 108:463–469

Khosrowshahi F, Arayici Y (2012) Roadmap for implementation of BIM in the UK construction industry. Eng Constr Architect Manag 19(6):610–635

McGraw Hill Construction (2009) The business value of BIM. Getting Building Information Modelling to the bottom line. https://bim.construction.com/research/pdfs/2009_BIM_SmartMarket_Report.pdf

McGraw Hill Construction (2010) The Business Value of BIM in Europe. Getting Building Information Modelling to the bottom line in the United Kingdom, France and Germany. http://images.autodesk.com/adsk/files/business_value_of_bim_in_europe_smr_final.pdf

Schlanbusch RD, Fufa SM, Häkkinen T, Vares S, Birgisdottir H, Ylmén P (2016) Experiences with LCA in the Nordic building industry: challenges, needs and solutions. Energy Procedia 96:82–93

Wong JKW, Zhou J (2015) Enhancing environmental sustainability over building life cycles through green BIM: a review. Autom Constr 57:156–165

Wu W, Issa RRA (2015) BIM execution planning in green building projects: LEED as a use case. J Manag Eng 31(1):1–18

Chapter 3
LCA Tools and BIM-Based LCA Methods to Support Decision-Making Process

Abstract Deepening LCA applications in construction sector, the LCA tools now available on the market and the BIM-based LCA methods are analysed to support environmental-oriented choices of building practitioners. In particular, the most established building LCA software are explored in terms of backgrounds, cost, assessment types, used databases, set of environmental indicators and further functionalities. Moreover, since LCA-BIM integration is growing in interest to steer building sector towards environmental sustainability, the BIM-based LCA methods are examined to suggest possible workflow according to data input, data analysis and data output, for endorsing the holistic approach of buildings, optimizing information exchange and time effort. The aim is to provide practitioners new ways to assess environmental impacts but also beyond, including energy analysis, cost evaluation and green building certifications.

3.1 The Plurality of LCA Tools for Buildings

In recent years, the introduction of regulatory systems and voluntary Green Building Rating Systems, aimed at considering the environmental performance of design choices throughout the life cycle, stresses the issue of tackling the environmental profile of buildings with reference to LCA methodology. However, unlike other industrial products, buildings represent extremely complex systems, not replicable and strictly related to the geographic location, climatic conditions, project requirements and ways of use. For this reason, as attested by LCA practitioners (see Chap. 2), the development of LCA studies is a complex and time-consuming activity. To support the design firms and supply chain stakeholders, various LCA tools and software have been developed over time in order to simplify the calculation process and facilitating LCA application in building practice. In this way, environmental efficiency becomes a further criteria for the decision-making process, with applications at different scales: from material, component and system level to whole building, district and city level.

On the international scene, a wide range of LCA software are currently available (Gantner et al. 2012; Peuportier and Putzeys 2005; Bayer et al. 2010; Annex31 2004; Peuportier et al. 2011; Quinones 2011; Hitchcock et al. 2011; Lehtinen et al.

A. Dalla Valle, *Change Management Towards Life Cycle AE(C) Practice*,
PoliMI SpringerBriefs, https://doi.org/10.1007/978-3-030-69981-9_3

2011). Nevertheless, these tools are generally used in practice only at the end of the process, when buildings are detailed in all construction materials and relative quantities are defined. Hence, the intent to affect the decision-making process is greatly reduced, since only in few cases they are used as tools for comparing different alternatives and for supporting project decisions. To facilitate the implementation of LCA starting from the early design phases and thus to allow practitioners to make informed choices in terms of environmental impacts, LCA tools tend to simplify the process of evaluation based on EN 15804/15978 standards, developing user-friendly software to enlarge the target audience. The main LCA software users are designers, especially architects and engineers, but also producers and environmental consultants.

The typical working principle for building LCA tools is the inputs data entry by the user, including the project bill of quantity of construction materials and products (foreground system). Thereafter they are attributed to the environmental data derived from the LCI databases (background system), being added, weighed and normalized according to the selected LCIA method (EC-JRC 2012). In this way, LCA results are strongly influenced both by the embedded LCI database and LCIA method, even if in some cases software let to choose from a list of options, contributed to more flexible instruments. The issue is that developing LCA studies with the same inputs data and assumptions but using different software lead to different results and, consequently, to divergent conclusions (Speck et al. 2015). For this purpose, the challenge is to reach comparable results and unambiguous interpretations as well as to provide practitioners the necessary LCA knowledge and expertise.

3.2 Top Five Most Popular Building-Related LCA Tools

Given the complexity of construction sector and the close link with the surrounding environment, LCA tools for buildings generally refer to a national context to comply with local regulations and databases, even if in some cases conceived with a broader perspective. To deepen their functionalities, potentialities and criticalities, the main building LCA software established in Europe and in the USA are analysed, exploring in particular Legep software (Germany), Ecosoft (Austria), Elodie (France) and Athena (North America). The selection derived from the identification of the country at the forefront of LCA use in building sector and of the most widespread tools in their respective contexts, also because expressly mentioned within Green Building Rating Systems. In addition, a focus is made on Tally, since it represents today the most widespread BIM-oriented LCA tool.

Each software is analysed according to the following aspects. First, the reference context and so if it derives from top down initiatives and the underlying process. Second, the cost as index of product dissemination. Third, the level of deepening of LCA analyses and so if it allows to perform different assessments according to the process phase. Fourth, the resulted environmental indicators and the used databases, specifying if they include EPDs of specific products. Finally, further potentialities offered by the tool, such as the integration of building systems, the simultaneously evaluation of life cycle costs and the interoperability with other software. In this way, it is displayed the set of key features to be considered, on one hand, by end-users during the decision-making process to select the most suitable tool for their needs and, on the other, by software companies to start disclosing in a transparent way the peculiarities and potentialities of the offered programs.

Legep (Fig. 3.1) is a life cycle analysis tool promoted by DBU—Deutsche Bundesstiftung Umwelt that allows to assess the energy needs for heating, hot water, electricity and related cost, as well as to understand the materials input–output flows, in compliance with DIN EN ISO 14040–44 standards. In the software, the building can be modelled using preassembled elements (on a range of about 6.000 different technologies), containing all the connected data about cost, energy, materials flow and impact, or manually defining the technological elements, changing the components layers and the materials characteristics. The full license is available at high-cost category (over €5.000), but affordable prices are set for student and training version, although for one-year period and with restricted functionalities. The software allows to perform complete and screening LCA and encompasses different database. For the material inventory, it relies, on one side, on Ökobau.dat, a free German ministerial database that provides both generic and specific manufacturer data, and, on the other, on Ecoinvent, the world leading database for environmental information, including construction and other industrial sectors. Instead, life cycle costs are established by sirAdos database, updated every year with specific products costs and price for new constructions, renovations and maintenance operations. Results are expressed according to the environmental indicators set by LCA standards (EN 15804:2012; EN 15978:2011), including global warming potential, acidification, ozone photochemical formation, ozone depletion, eutrophication, primary energy consumption from renewable and non-renewable sources, with the chance to include partially or totally both indicators and life cycle stages. Legep is a design support tool that provides a holistic approach of the project, by the integration of five modules: costs, energy, Life Cycle Costs, Life Cycle Assessment, materials and health. Moreover, it is compatible with DGNB rating system, BNB Sustainability Certification as well as with GAEB standard for exchanging data between process practitioners.

Ecosoft (Fig. 3.2) is an LCA online software developed by IBO—Austrian Insti-

SOFTWARE	CONTEXT	MODELING	COST	ASSESSMENT	DATABASE	INDICATORS	INTEGRABILITY
analysed software	developer	building model definition	software cost	level of detail	database embedded	environmental indicators in output	further potential
			HIGH				INTEROPERABILITY
		AUTOMATIC	MEDIUM	COMPLETE LCA	INTERNATIONAL	COMPLETE	RATING SYSTEM
LEGEP	GERMANY	MANUAL	LOW	SIMPLIFIED LCA	NATIONAL	BASIC	LCC
	developed by DBU - Deutsche Bundesstiftung Umwelt	manual definition of technological elements or use of preassembled elements	high costs category (over € 5,000), student version and training version	double level of analysis, with the chance of performing separate calculation based on partial or total life cycle phases	possibility to use the international Ecoinvent database or the German Ökobaudat database and for costs the German database sirAdos	global warming, acidification, total primary energy consumption, ozone depletion, eutrophication, etc.	interoperable with GAEB standard and compatible with DGNB and BNB, functional to perform LCC evaluations

Fig. 3.1 Key features of Legep LCA software

SOFTWARE	CONTEXT	MODELING	COST	ASSESSMENT	DATABASE	INDICATORS	INTEGRABILITY
analysed software	developer	building model definition	software cost	level of detail	database embedded	environmental indicators in output	further potential
			HIGH				INTEROPERABILITY
		AUTOMATIC	MEDIUM	COMPLETE LCA	INTERNATIONAL	COMPLETE	RATING SYSTEM
ECOSOFT	AUSTRIA	MANUAL	LOW	SIMPLIFIED LCA	NATIONAL	BASIC	LCC
	developed by IBO - Österreichisches Institut für Baubiologie and promoted by the government	manual definition of technological elements	low costs category (less than € 500), educational version and free test	possibility to perform separate calculation based on partial or total life cycle phases	Austrian IBO database (adjustment of Ecoinvent database)	global warming, acidification and primary non-renewable energy consumption	integration in the near future

Fig. 3.2 Key features of Ecosoft LCA software

tute for Healthy and Ecological Building since 1998 and promoted by the Austrian government, which provides funding for residential buildings that demonstrate low environmental impacts. Based on Microsoft Excel, it allows users the breakdown of buildings in components and layers, defining for each the thickness and share of the embedded materials, later assigned to the environmental profile of database, with the chance to modify default material density and durability. The full version of the software is part of low-cost category (less than €500), offering also educational version and free test, to edit and assess buildings right after the login. The software allows to conduct a single type of LCA analysis, considering as appropriate the entire life cycle or specific stages. The reference IBO database (Ecoinvent version suitable for Austrian context) is constantly updated and enriched by the free digital platform Baubook, which contains a wide range of products and EPDs, enabling manufacturers to perform self-certifications. Three environmental indicators are covered: global warming potential, acidification and non-renewable primary energy, by which the ecological index of building constructions is calculated ranging from 0 to 100 points. In the next future, both other databases and environmental indicators are expected to be implemented, in order to ensure greater choice for building practitioners. Ecosoft enables the assessment of the environmental performance of buildings, including HVAC systems, allowing the exportation of the model into other programs, such as energy simulation software. However, the main weakness of the tool is the failure to perform LCC studies and the lack of interoperability with CAD and BIM tools.

Elodie (Fig. 3.3) is a tool for the environmental assessment of buildings, devel-

SOFTWARE	CONTEXT	MODELING	COST	ASSESSMENT	DATABASE	INDICATORS	INTEGRABILITY
analysed software	developer	building model definition	software cost	level of detail	database embedded	environmental indicators in output	further potential
			HIGH				INTEROPERABILITY
		AUTOMATIC	MEDIUM	COMPLETE LCA	INTERNATIONAL	COMPLETE	RATING SYSTEM
ELODIE	FRANCE	MANUAL	LOW	SIMPLIFIED LCA	NATIONAL	BASIC	LCC
	developed by CSTB - Centre Scientifique et Technique du Bâtiment	BIM model import or manual definition of technological elements	medium costs category (price between € 500 and € 5,000), beta version available online for free	double level of analysis, with the chance of performing separate calculation based on partial or total life cycle phases	French INIES database	global warming, acidification, total primary energy consumption, ozone depletion, eutrophication, etc.	interoperable with BIM models in IFC format and compatible with HQE Performance, functional to perform LCC evaluations

Fig. 3.3 Key features of Elodie LCA software

SOFTWARE	CONTEXT	MODELING	COST	ASSESSMENT	DATABASE	INDICATORS	INTEGRABILITY
analysed software	developer	building model definition	software cost	level of detail	database embedded	environmental indicators in output	further potential
			HIGH				INTEROPERABILITY
		AUTOMATIC	MEDIUM	COMPLETE LCA	INTERNATIONAL	COMPLETE	RATING SYSTEM
ATHENA	NORTH AMERICA	MANUAL	LOW	SIMPLIFIED LCA	NATIONAL	BASIC	LCC
	developed by the Athena Sustainable Materials Institute	manual definition of technological elements or use of preassembled elements	all Athena tools are available for free	double level of analysis, with the chance of performing separate calculation based on partial or total life cycle phases	Athena datasets and US Life Cycle Inventory Database	global warming, acidification, total primary energy consumption, ozone depletion, eutrophication, etc.	compatible with LEED, possibility to import bill of materials from any CAD program

Fig. 3.4 Key features of Athena LCA software

oped by CSTB—Centre Scientifique et Technique du Bâtiment and distributed after a testing phase with a mandatory training course for its use. The software, based on a modular structure, enables the calculation of the building environmental profile during the entire life cycle and discloses additional information on acoustic comfort, indoor air quality and overall construction costs. As always, users define the building into the set of technological elements, matching them with the related material datasets, although the full version permits to directly import the BIM building model. The full license, composed of Elodie and even BIM, is available at medium-cost category, with varying price according to the number of users, while the beta version is accessible online for free. Simplified and complete LCA studies are included within the software, based on the internal INIES database, which consist both of building materials EPDs (called FDES in French) and electrical and HVAC components PEPs (Product Environmental Profile). The resulting environmental indicators are fifteen, in accordance with EN 15978 standard. Tool potentialities concern the join assessment of LCA analysis with acoustic, air quality and cost analysis, meeting HQE Performance requirements. Moreover, another crucial aspect is the interoperability with BIM through IFC format, allowing practitioners to save time and increase productivity, by-passing construction material quantity calculations.

Athena (Fig. 3.4) is a user-friendly software developed by Athena Sustainable Materials Institute for comparing the environmental profile of alternate design scenarios of both construction assemblies and whole buildings. In particular, it provides two different tools: Athena Impact Estimator for Buildings and Athena EcoCalculator for Assemblies, distinguished by different level of detail for suiting from beginning to end the design process. Athena EcoCalculator is a spreadsheet tool, very intuitive and fast but limited in options since characterized by predefined technological configurations, not editable. Users need only to input the square footage of the selected default assembly to obtain the life cycle results, except operational energy. By contrast, Athena Impact Estimator is a stand-alone program, more flexible and accurate, in which users model customized assemblies and envelope configurations and input energy consumption. All Athena tools are available for free and are based on own Athena datasets and US Life Cycle Inventory database, referring thus to USA context. These databases are regionally sensitive, taking into consideration the differences in manufacturing technologies, transportations, electricity grids and products recycled content, and they include continually updated cradle-to-grave EPDs.

SOFTWARE	CONTEXT	MODELING	COST	ASSESSMENT	DATABASE	INDICATORS	INTEGRABILITY
analysed software	developer	building model definition	software cost	level of detail	database embedded	environmental indicators in output	further potential
			HIGH				INTEROPERABILITY
		AUTOMATIC	MEDIUM	COMPLETE LCA	INTERNATIONAL	COMPLETE	RATING SYSTEM
TALLY	NORTH AMERICA	MANUAL	LOW	SIMPLIFIED LCA	NATIONAL	BASIC	LCC
	developed by KT Innovations, Autodesk and Thinkstep	manual definition of technological elements within the BIM model	medium costs category (between € 500 and € 5,000), 10-day trial and academic version	double level of analysis, with the chance of performing separate calculation based on partial or total life cycle phases	custom designed LCA database with data from GaBi databases	global warming, acidification, total primary energy consumption, ozone depletion, eutrophication, etc.	integrated in BIM models and compatible with LEED

Fig. 3.5 Key features of Tally BIM-LCA app

Moreover, in both cases report indicators are consistent with the latest US TRACI methodology: global warming, acidification, human health respiratory effects, ozone depletion, smog and eutrophication potential, adding in Athena Impact Estimator also fossil fuel consumption. Athena tools are not BIM interoperable and only in Athena Impact Estimator it is possible to import bill of materials from CAD program. Note that Athena Impact Estimator is expressly mentioned in the LCA credit of LEED v4 as a reference tool for North America context, promoting therefore its dissemination.

Tally (Fig. 3.5) is the first LCA-BIM app, launched by KT Innovations, an affiliate of Kieran Timberlake, and then supported by Autodesk Sustainability Solutions and Thinkstep for providing easy-to-use tool that brings accurate life cycle data throughout the design process. In this way, working on Revit, construction industry professionals directly associate BIM objects with the environmental material datasets embedded in Tally, with a greater impact on the decision-making process. A 10-day trial is available to start approach the tool and the one-year full version falls within medium-cost category if for commercial license, while it is free of charge at discretion of KT Innovations if used for non-commercial, educational and research purposes. The software allows users to perform during the design process different types of LCA analysis with different level of detail, including materials production, maintenance, replacement and end-of-life. In addition, practitioners are enabled to adjust as appropriate some reference values, such as transport distances between manufacturers and building site, the planned replacement cycles and the operational energy consumption. The tool utilizes a custom designed LCA database that combines material attributes, assembly details and construction specifications with the environmental impacts data retrieved from GaBi database. Tally reports environmental impacts according to TRACI methodology, because adopted by LEED certification and accepted by Living Building Challenge certification program. In particular, LCA results concern: ozone depletion, global climate, acidification, eutrophication, smog formation, human health particulate, human health cancer, human health non-cancer, ecotoxicity and primary energy demand. Since LEED-compliant, Tally is commonly used to earn LEED v4 Building Life Cycle Impact Reduction credit but also to examine building materials and components, excluding mechanical equipment, because typically considered out of scope and not embedding cost information. Nevertheless, Tally has the potential to integrate LCA and BIM, encouraging LCA application during the design process for better environmental decision-making.

3.3 BIM-Based LCA Methods and Beyond for Buildings

LCA-BIM integration is growing in interest to steer building sector towards environmental sustainability. Indeed, on one hand, LCA is widely recognized as the most complete methods to assess the environmental impacts of buildings over its life cycle. On the other, BIM is increasingly adopted by building practitioners to share graphical and non-graphical information of building projects from inception onward. In this way, LCA-BIM integration allows to achieve a holistic overview by providing feedback on the environmental impacts of building design choices, solving some redundant and time-consuming tasks typical of building LCA studies.

In this context, many LCA-BIM tools are developed as standalone software or as plug-in, as in the case of Tally, to quantify the environmental LCA and provide comparative analysis of design options. To suggest the workflow among several tools, a review of BIM-based LCA methods for buildings is carried out, focusing in accordance with the main steps required for LCA design on: data input, data analysis and data output (Seo et al. 2007; Soust-Verdaguer et al. 2017).

As well known, data inventory represents the most time-consuming phase for building LCA application and turns out to be considerably relieved in BIM-based LCA methods. Indeed, the key issue of LCA-BIM integration is that the bill of materials is directly calculated and extracted from BIM model, increasing the accuracy and consistency of quantitative data. To this end, the Level of Development (LOD) of BIM model plays a crucial role, because the related level of detail is linked with the phase of the design process, significantly affecting the automatic calculation of building components, materials and objects. Most studies do no declare the LOD of the adopted BIM model, but a couple of cases specifies that the LOD required for the developed methods is LOD 200 (Ajayi et al. 2015) and LOD 300 (Lee et al. 2015).

Concerning input data, all BIM-based LCA methods assume as functional unit of the assessment the whole building, apart from few cases that consider 1m2 of heated area (Houlihan et al. 2014). The buildings under investigation refer always to new construction, except one case focused on existing building (Peng 2016), and encompass different building typologies, including office, residential, school and university facilities. Despite the recommended LCA system boundaries (EN 15978:2011), the studies generally focus on product stage and in part on use stage, excluding completely construction and end of life stages. In this way, it emerges how LCA studies currently involve only the stages easily modelled, slightly including the scenario stages that entail temporal aspects, such as maintenance and replacement, and completely excluding recycling and reuse. Moreover, the analysed BIM-based LCA methods adopted various environmental databases, with different level of accuracy and validity. They regard, on one side, the internationally recognized Ecoinvent database, constantly updated and containing general but also product specific data (Houlihan et al. 2014) and, on the other, ICE database that provide standard data on embodied energy and embodied carbon of the principal construction materials (Iddon and Firth 2013; Peng 2016). By contrast, regional databases are employed, as Athena database for USA and Canadian context (Ajayi et al. 2015; Jrade and Jalaei

2013; Jalaei and Jrade 2014; Iddon and Firth 2013) and Korea Life Cycle Inventory for Korean context (Lee et al. 2015; Shin and Cho 2015).

According to the second step of life cycle design, data analysis involves the process of transfer and interconnection of the data gathered by BIM model, the energy simulation software and the LCA software used for the calculations, aiming at interoperability solutions. All methods use Autodesk Revit to develop the BIM model, except for few cases that select DProfiler (Basbagill et al. 2013) and ArchiCAD (Shin and Cho 2015). Energy simulations are performed through a wide range of software, such as: Green Building Studio, eQUEST, Simien, Autodesk Ecotect, Integrated Environmental Solutions and EcoDesigner. Lastly, the LCA tools adopted are Athena Impact Estimator (Ajayi et al. 2015; Jalaei and Jrade 2014) and SimaPro (Houlihan et al. 2014), being in some case also jointly used (Basbagill et al. 2013).

Despite the goals heterogeneity of studies, the data exchange procedure is shared, exploiting BIM for the automatic bill of materials quantities, which is afterward exported to Excel to manually adjust and manage data before LCA application. This proves how the match between the quantitative data extracted by BIM model and the environmental data embedded within LCA tools is still not fully automated, requiring hand-operated links. For this purpose, to avoid double work and automatize data exchange between BIM and LCA, it is important to organize from the outset in a systematic way the BIM model, classifying building objects according to recognized classification systems, like UniFormat and Masterformat, but also components materials according to the LCA database nomenclature intended to use.

Finally, output data focus especially on CO_2 emissions, resulting screening LCA studies (Gantner et al. 2012), since they mainly assess product and use stages considering one environmental indicator. In addition, sensitivity analysis is developed only in few cases, on one side, to search minimum and maximum embodied impacts for each building components (Basbagill et al. 2013) and, on the other, to identify the parameters that have the greatest impact on building performance (Peng 2016). Regardless of the level of detail, however, it is recognized the importance of applying the proposed BIM-based LCA methods starting from the early design phases to improve performance in construction industry (Basbagill et al. 2013; Li et al. 2012; Jalaei and Jrade 2013, 2014; Means and Guggemos 2015; Rock et al. 2018).

Note that the challenge of LCA-BIM integration is faced in different ways among the case studies, excluding LCA-BIM plug-in as Tally that actually represent the more effective method. If LCA tools are not BIM interoperable, it is suggested a manual process via Excel or an automatic process by a bi-directional data exchange and synchronization (Kovacic 2016). If LCA tools are BIM interoperable, data are automatically extracted from BIM model via IFC format. Alternatively, environmental LCA data are directly included into BIM object proprieties (Anton and Diaz 2014). In this way, data-handling with Excel software turns out to be the easiest way, even if the lack of automated synchronization prone to errors due to model-changes. The use of a bi-directional data exchange, as BuildingOne, requires a lot of effort and experience to customize the database. Instead, the use of BIM interoperable LCA tools avoids manual data re-entry, facilitating the comparison among different alternatives and allowing to perform real-time assessment. Nevertheless, even in this case, any BIM model changes entail the re-importing of the model into the LCA platform, still not fully interoperable. Lastly, the inclusion of the environmental LCA data in the set of BIM objects information allows to include environmental sustainability as additional design criteria within the decision-making process.

However, in line with current trends (see Chap. 2), building LCA methods are not only correlated with BIM but also with Green BIM towards a holistic approach to sustainability. To this end, follow-up methods have been developed over time to help practitioner in further exploiting BIM functionalities to develop LCA studies in conjunction with the most varied design analyses. Among the almost infinite range of possibilities, some authors propose an automated BIM-based method to match LCA studies with energy analysis, lighting simulation, cost evaluations and Green Building Rating Systems certifications (Jalaei and Jrade 2013, 2014). Others focus on the information needed to conduct LCA and LCC in order to jointly obtain them in an efficient and accurate manner using BIM tools, enhancing thus during the decision-making process both LCA and LCC implementation (Shin and Cho 2015). Going beyond environmental and economic issues, some methods pursue automatic assessments of building sustainability life cycle, taking advantage of all BIM dimensions, in particular 4D schedule, 5D cost and 6D sustainability (Yung and Wang 2014).

In the construction sector, the application of BIM-based LCA methods and beyond has the potential to change the traditional design practice and to encourage the development of high-performance buildings. However, it is worth underlining that all these methods are proposed by the scientific community to gradually reduce the environmental, economic and social impacts of buildings. They are hence not necessary adopted by designers and practitioners, demanding an in-depth analysis of current practice for improving automatic process and data exchange and promoting an integrated planning and life cycle approach to the built environment.

References

Ajayi SO, Oyedele LO, Ceranic B, Gallanagh M, Kadiri KO (2015) Life cycle environmental performance of material specification: a BIM-enhanced comparative assessment. Int J Sustain Build Technol Urban Dev 6:14–24

Annex31 (2004) Directory of tools: a Survey of LCA Tools. Assessment frameworks, rating systems, technical guidelines, catalogues, checklists and certificates

Anton LA, Diaz J (2014) Integration of life cycle assessment in a BIM environment. Procedia Eng 85:26–32

Basbagill J, Flager F, Lepech M, Fischer M (2013) Application of life-cycle assessment to early stage building design for reduced embodied environmental impacts. Build Environ 60:81–92

Bayer C, Gamble M, Gentry R, Joshi S (2010) AIA Guide to building Life Cycle Assessment in practice. https://www.aia.org/aiaucmp/groups/aia/documents/pdf/aiab082942.pdf

EN 15978:2011. Sustainability of construction works. Assessment of environmental performance of buildings. Calculation methods.

European Commission—Joint Research Centre (2012) ILCD Handbook. Towards more sus- tainable production and consumption for a resource efficient Europe. Publications Office of the European Union: Luxembourg
Gantner J, Saunders T, Lasvaux S (2012) EeBguide—Requirements for building LCA tool designer. Operational guidance for Life Cycle Assessment studies of the Energy Efficient Buildings Initiative
Hitchcock D, Schenk R, Gordy T (2011) Directory of sustainability Life Cycle Assessment tools. International Society of Sustainability Professionals. https://www.sustainabilityprofessionals.org/system/files/ISSP_Life_Cycle_Assessment_Tools_Directory.pdf
Houlihan A, Georges L, Dokka TH, Haase M, Time B, Lien AG et al (2014) A net zero emission concept analysis of a single-family house. Energy Build 74:101–110
Iddon CR, Firth SK (2013) Embodied and operational energy for new-build housing: a case study of construction methods in the UK. Energy Build 67:479–488
Jalaei F, Jrade A (2014) An automated BIM model to conceptually design, analyze, simulate and assess sustainable building projects. J Constru Eng, 1–21
Jrade A, Jalaei F (2013) Integrating building information modelling with sustainability to design building projects at the conceptual stage. Build Simul 6:429–444
Kovacic I (2016) BIM for life cycle assessment—a case study: Redensification. In WBC16 Understanding impacts and functioning of different solutions, 1–16
Lee S, Tae S, Roh S, Kim T (2015) Green template for life cycle assessment of buildings based on Building Information modelling: focus on embodied environmental impact. Sustainability 7:16498–16512
Lehtinen H, Saarentaus A, Rouhiainen J, Pitts M, Azapagic A (2011) A Review of LCA methods and tools and their suitability for SMEs
Li B, Fu FF, Zhong H, Luo HB (2012) Research on the computational model for carbon emissions in building construction stage based on BIM. Struct Surv 30(5):411–425
Means P, Guggemos A (2015) Framework for Life Cycle Assessment (LCA) based environmental decision making during the conceptual design phase for commercial buildings. Procedia Eng 118:802–812
Peng C (2016) Calculation of a building's life cycle carbon emissions based on Ecotect and building information modelling. J Cleaner Prod 112:453–465
Peuportier B, Putzeys K (2005) PRESCO—Inter-comparison and benchmarking of LCA-based Environmental Assessment and Design Tools
Peuportier B, Scarpellini S, Glaumann M, Malmqvist T, Krigsvol G, Wetzel C et al (2011) ENSLIC building: state of the art report and collection of published material
Quinones MC (2011) Decision support system for building construction product selection using Life-Cycle Management (LCM)
Röck M, Hollberg A, Habert G, Passer A (2018) LCA and BIM: integrated assessment and visualization of building elements' embodied impacts for design guidance in early stages. Procedia CIRP 69:218–223
Seo S, Tucker S, Newton P (2007) Automated material selection and environmental assessment in the context of 3D building modelling. J Green Build 2(2):51–61
Shin Y, Cho K (2015) BIM application to select appropriate design alternative with consideration of LCA and LCCA. Math Problems Eng, 1–15
Soust-Verdaguer B, Llatas C, García-Martínez A (2017) Critical review of BIM-based LCA method to buildings. Energy Build 136:110–120

Speck R, Selke S, Auras R, Fitzsimmons J (2015) Life cycle assessment software: selection can impact results. J Ind Ecol 20(1):18–28

Yung P, Wang X (2014) A 6D CAD model for the automatic assessment of building sustainability. Int J Adv Rob Syst 11(131):1–10

Chapter 4
Building-Related LCA Application Review Within Construction Sector

Abstract For fostering the development of LCA studies in building practice, a sample of building-related LCA applications are reviewed, focusing on the European context and the time period between 2010 to date. In particular, the large ensemble of quantitative life cycle information (foreground system) are break it down since covered by designers, rather than the environmental information (background system) referred to specific databases and/or certifications. The aim is to verify the completeness and quality of the now considered inventory, on one hand, for providing the reference state for future assessments and, on the other, for highlighting the missing information to be overcome. Indeed, the consistency and types of the accounted quantitative information greatly affects the level of completeness of the study, while the quality of such information highly influence the reliability and accuracy of the results. To this end, the analysis gives prominence to possible sources for data collection and areas for improvement.

4.1 LCA Studies Completeness and Information Quality

Despite the shared standards and action plans and the wide range of LCA tools now available, life cycle approach is still not established in design and construction practice, revealing uneven and fragmented LCA applications.

The major constraint is the large amount of information required for developing building-related LCA studies, compounded by the fact that most data are generally defined only at advanced phase of the process, including the type of materials and the quantities of all construction components. In addition, unlike other industrial sectors where products are manufactured in series, with industrialized processes and short-term lifetime, buildings are extremely complex systems, unique, not replicable and strongly influenced by the context, distinguished by an increasing fragmentation of activities, by a multitude of actors and by long life cycles. Therefore, if in other fields LCA methodology can be applied to finished products that can be improved and place again on the market, in construction sector this is not possible. Ad hoc studies must be performed for each building in order to affect the decision-making process over the entire life cycle, taking into account the specificity of the construction in question.

A. Dalla Valle, *Change Management Towards Life Cycle AE(C) Practice*,
PoliMI SpringerBriefs, https://doi.org/10.1007/978-3-030-69981-9_4

Moreover, beyond the peculiarities of building sector, the same LCA methodology has unresolved aspects to be overcome, enclosing the availability of environmental data and the comparability of results.

All these factors strongly influence the development and the level of detail of the life cycle inventory, widely recognized as the most tricky and demanding phase of LCA studies. The complete list of information to be considered during the building assessment is provided by the reference standard (EN 15978:2011), calling for each two distinct types of data: the quantitative data and the environmental data. For instance, the kilos of concrete used in the building (quantitative data) are associated with the respective emissions, raw materials extracted, etc. (environmental data), that is the elementary flows in input and output between the technosphere and the ecosphere. The consistency and types of the gathered quantitative information greatly affects the level of completeness of the study, while the quality of such information highly influence the reliability and accuracy of results.

For encouraging the development of LCA studies in building practice, it is important to improve and support the implementation of the life cycle inventory, giving priority especially to the quantitative life cycle information covered by practitioners and designers (foreground system), rather than on the environmental information referred to specific databases and/or certifications (background system). To this end, a literature review was conducted on a sample of building-related LCA studies conducted within the European context from 2010 to date. The focus is on the quantitative information currently assessed or, alternatively, excluded from the evaluations for pursuing a twofold purpose. On one side, the aim is to survey current LCA applications and verify the completeness and quality of the now considered inventory information. On the other, the goal is to provide an overview of the set of primary information and related sources for offering data collection hints and references in order to support future assessments.

4.2 Deepening LCA Foreground Information Stage-by-Stage

The quantitative information required for building-related LCA studies are explained for each stage of life cycle, pointing out the connected types and quality, from the more virtuous to the less upstanding case. This is established on the basis of the source used for retrieving the information in object, taking into account the degree of uncertainty. High level is thus assigned for primary data monitored in the field and decreasing level respectively for data calculated from design documentations, taken from statistical data and assumed from literature. Furthermore, the information currently excluded from the LCA studies are highlighted in order to suggest possible areas for improvement.

The product stage (Fig. 4.1) is systematically included within building-related LCA studies, jointly encompassing the extraction of raw materials (A1), the transport

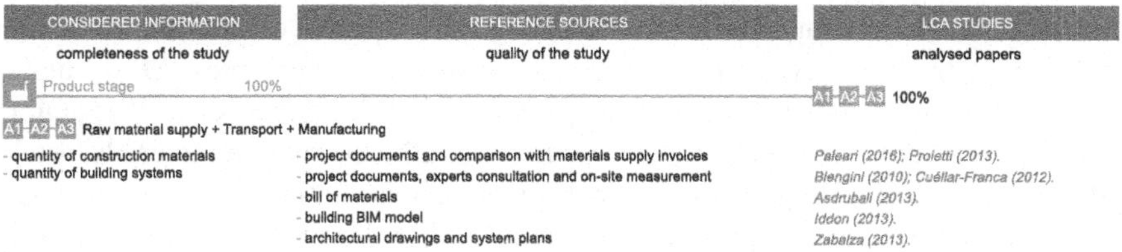

Fig. 4.1 Information and sources considered for the product stage

from the extraction site to the factory (A2) and the manufacturing process of building products and components (A3). It consists basically of the quantities of construction materials and, in some cases, also of the adopted building systems.

Some authors (Paleari et al. 2016; Proietti et al. 2013) determine materials quantities from project documents and compare the calculated volumes with the invoices paid for supplying the materials. Instead, other authors (Blengini and Di Carlo 2010; Cuéllar-Franca and Azapagic 2012) define them based on project drawings and technical data sheets, checking the amount of materials through experts consultation and on-site measurements. Finally, few studies assume the quantities derived from the bill of materials (Asdrubali et al. 2013), automatically extrapolated from the building BIM model (Iddon et al. 2013) or manually calculated from architectural and engineering drawings (Zabalza et al. 2013).

The construction stage (Fig. 4.2) is included in 85% of cases and excluded in the remaining 15%. If included, 18% of studies consider only the transport from factory to building site (A4), 17% just the construction and installation process (A5) and 65% of studies both modules.

Concerning the transport phase (A4), LCA studies assess the transport of construction materials from manufacturing plant to construction site. Some authors (Paleari et al. 2016; Proietti et al. 2013) define the distances, as well as the means of transport with the related maximum capacity, from the set of materials procurement invoices, shipping notes and direct contacts with suppliers. In addition, Paleari et al. (2016) specify that the distances are calculated from the first production sites to the building site encompassing, when necessary, all the intermediate processing steps. By contrast, other authors (Blengini and Di Carlo 2010; Dahlstrøm et al. 2012) determine the

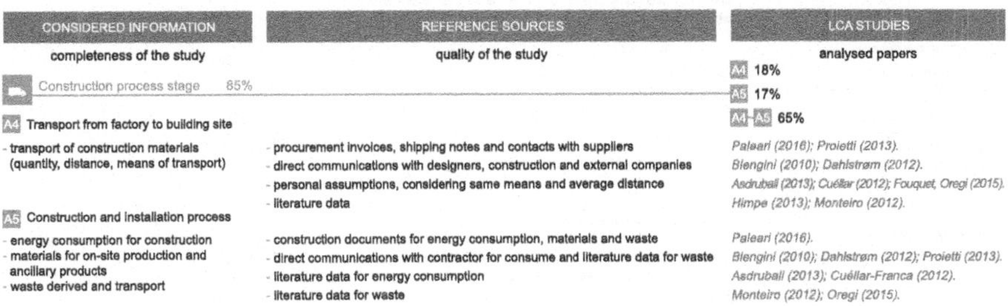

Fig. 4.2 Information and sources considered for the construction process stage

transport of construction materials and, although not required, of workers, collecting information by interviewing designers and construction companies. Most studies (Asdrubali et al. 2013; Cuéllar-Franca and Azapagic 2012; Fouquet et al. 2015; Oregi et al. 2015) assume the same transport by road for all construction materials, establishing a common average distance or different representative range: at province, region or country levels. Lastly, the remaining studies derive this information from literature data relative to the specific reference country (Himpe et al. 2013) or to the European context (Monteiro and Freire 2012).

In the construction process phase (A5), the energy consumption for construction activities and the derived construction waste are mainly considered. In particular, Paleari et al. (2016) includes the energy and water used during operation by equipment, the materials used for on-site production, the auxiliary products, the protective materials and the disposal of construction waste, retrieving real data from construction documents. Instead, others (Blengini and Di Carlo 2010; Dahlstrøm et al. 2012; Proietti et al. 2013) gather the energy used for construction and installation processes through direct communications with contractors but consider for construction waste the average percentage set by literature. Finally, some authors assess based on literature data only the energy consumption of construction process (Asdrubali et al. 2013; Cuéllar-Franca and Azapagic 2012) or just the amount of construction waste and the relative transport to landfill (Monteiro and Freire 2012; Oregi et al. 2015).

The use stage (Fig. 4.3) is included in all examined LCA studies, even if the assessments are limited to some modules: 100% of cases consider the operational energy consumption (B6), 65% the maintenance and replacement operations (B2 and B4), 32% the operational water consumption (B7), while use, repair and refurbishment modules are excluded (B1, B3 and B5).

Maintenance and replacement processes (B2 and B4) are usually joined together. Oregi et al. (2015) assume the service life of building materials according to the connected EPDs specifications or, where not available, from maintenance industry

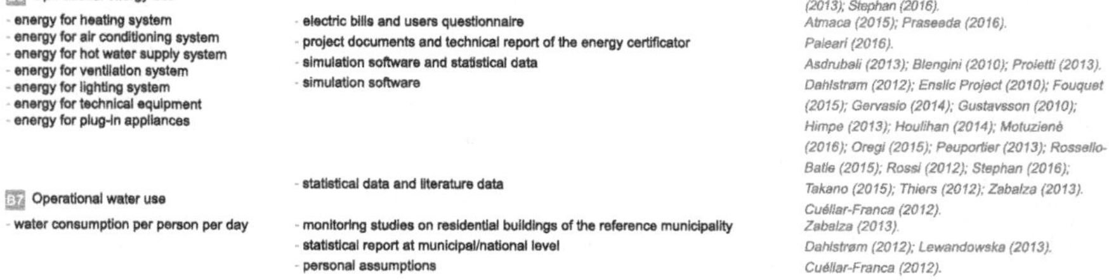

Fig. 4.3 Information and sources considered for the use stage

databases, including products transport. Paleari et al. (2016) define the lifespan of the main building components based on manufacturer data and by experience, comparing the collected values with durability reports and scenarios developed by tenants and owners. Instead, other authors (Blengini and Di Carlo 2010; Monteiro and Freire 2012) trace this information by questioning designers, builders and manufacturers and, when necessary, by literature. Lastly, the majority of studies (Atmaca and Atmaca 2015; Cellura et al. 2014; Cuéllar-Franca and Azapagi 2012; Dahlstrøm et al. 2012; Himpe et al. 2013; Proietti et al. 2013; Stephan and Stephan 2016) takes for the different building elements the literature maintenance and replacement cycles.

The operational energy use (B6) covers fully or partly the following contributions: heating, cooling, domestic hot water, ventilation, lighting system as well as technical equipment and plug-in appliances. Some authors (Atmaca and Atmaca 2015; Praseeda et al. 2016) gather the energy consumption of buildings by electric bills and questionnaires to users, while others (Paleari et al. 2016) infer them by project documents and energy certifications. By contrast, sometimes (Asdrubali et al. 2013; Blengini and Di Carlo 2010; Proietti et al. 2013) consumption are segregated according to their reliance on building, such as energy for heating, cooling, ventilation and domestic hot water that are calculated through digital simulations, or alternatively on users addiction, like electrical appliances, cooking, washing and lighting systems that are sourced from statistics. Nevertheless, most studies (Dahlstrøm et al. 2012; Enslic Project 2010; Fouquet et al. 2015; Gervasio et al. 2014; Gustavsson and Joelsson 2010; Himpe et al. 2013; Houlihan et al. 2014; Motuzienè et al. 2016; Oregi et al. 2015; Peuportier et al. 2013; Rossello-Batle et al. 2015; Rossi et al. 2012; Stephan and Stephan 2016; Takano et al. 2015; Thiers and Peuportier 2012; Zabalza et al. 2013) assess the energy consumption by performing energy simulations, using software adapted to specific contexts, such as COMFIE for France, SIMIEN for Norway, CALENER VYP for Spain, EPB for Belgium, or general modelling tools, such as DesignBuilder. Lastly, the whole domestic energy consumption is retrieved from statistical data and then broken down into heating, domestic water, lighting and appliances (Cuéllar-Franca and Azapagic 2012).

Regarding the operational water use (B7), it is rarely considered into building-related LCA studies. Zabalza et al. (2013) evaluate it on the basis of municipality monitoring on residential buildings, while Dahlstrøm et al. (2012) based on statistical reports carried out at municipal level and Lewandowska et al. (2013) on reports developed at national level. Finally, water consumption are included by Cuéllar-Franca and Azapagic (2012) simply assuming the average water consumption per person per day.

The end of life stage (Fig. 4.4) is counted in 45% of studies, of which 67% of cases consider the deconstruction and demolition processes (C1), 100% the transport from building site to landfill (C2) and 86% the waste processing (C3) and disposal (C4).

Some authors (Blengini and Di Carlo 2010; Cuéllar- Franca and Azapagic 2012; Himpe et al. 2013; Takano et al. 2015) use literature data for deconstruction and demolition operations as well as for the different end-of-life scenarios. Instead,

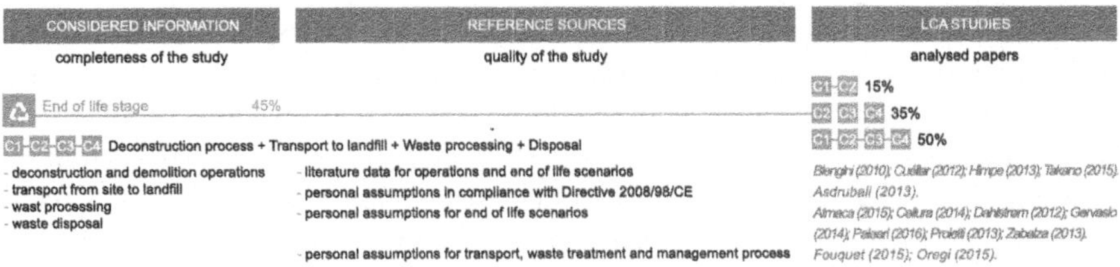

Fig. 4.4 Information and sources considered for the end of life stage

others establish the material scenarios choosing between landfill treatment or recycling facilities, in order to comply with Directive 2008/98/ EC and provide thus the recovery of 70% of weight of all waste products. In the majority of studies (Atmaca and Atmaca 2015; Cellura et al. 2014; Dahlstrøm et al. 2012; Gervasio et al. 2014; Paleari et al. 2016; Proietti et al. 2013; Zabalza et al. 2013) assumptions are made for each material, considering for instance landfill for wood, bitumen and polystyrene, and recycling for steel, aluminium and plastic elements. Lastly, few authors (Fouquet et al. 2015; Oregi et al. 2015) take into account an average distance between the building site and the disposal units, including also the waste treatment and management process.

4.3 Missing LCA Foreground Information to Be Filled in

Reviewing building-related LCA applications, it emerges how the now available LCA studies derives from the academic world and, to a lesser extent, from the professional world, since in the latter they generally remain unreleased works for preserving the confidentiality of the commercial, industrial and environmental information. Consequently, it is important to stress that the examined LCA studies were carried out ex-post by researchers for research purposes, without affecting the decision-making related to design, construction and management processes. Performing the assessment at the end of the process, as generally happens when building is complete, certainly helps to figure out its environmental profile but does not allow to orient design choices and to effectively reduce construction impacts. In this way, putting the spotlight on the phase of the process in which LCA studies are developed is pivotal not only to predetermine the completeness and quality of the retrieved information, but also to ascertain the effect on the decision-making process.

However, it is worth mentioning that the building-related LCA review provides assistance to implement LCA in practice also within architectural, engineering and construction companies, as main actors responsible for the built environment. Indeed, it suggests the principal information to be considered as well as a set of possible sources and methodological references valuable to perform future LCA studies and to improve the reliability and accuracy of results. Moreover, the information currently excluded from LCA studies advices possible areas for improvement to enhance the information demanded for the foreground system and thus to strengthen a more holistic perspective of the construction sector.

The product stage (A1-A3) typically fails to address the adopted building systems, risk overlooking the impacts derived from the manufacturing of high technological components and systems in favour of the reducing electricity consumption benefits during the use phase. Concerning the construction process stage, the information currently excluded for the transport module (A4) concerns the packaging, the set of materials lost or damaged during carriage and the transfer of construction machinery and equipment. Whereas for the construction and installation process (A5), they refer to the energy used for products storage and for the building site facilities, to the transport of materials, waste and equipment within the site, to the water used for on-site production and cleaning as well as for machinery.

The use stage consists in a series of modules that ranges from operational activities to energy and water consumption. The use module (B1) and the repair and restoration processes (B3 and B5, respectively) are always excluded by LCA studies. In the first case, because the information is highly specific, by calling for the quantification of substances emitted from facade, roofing and cladding surfaces. Conversely, in the second case, because the required operations cannot be defined in a preventive manner, but only after a damage. Instead, the maintenance and replacement operations (B2 and B4, respectively) are often considered in terms of materials, omitting the water and energy used for works, the waste produced with the related transport to landfill and waste management process. With regard to the operational energy use (B6), the module transversally takes into account the energy used for pumps, automation and other building-integrated systems, such as lifts, escalators, security and communication systems. Finally, by providing brief description on both the operational water use (B7) and end of life stage (C1-C4), the LCA studies not allow to define the contributions currently neglected from these modules.

For the shake of completeness and to avoid shifting problems, in the next future it is recommended to include in building-related LCA studies the now omitted information. Nevertheless, it is important to stress that since the missing data are in a great deal of detail, they must be trace with purposeful intent, bearing that in mind from the outset of the process in order to monitor and track them towards sustainability.

References

Asdrubali F, Baldassarri C, Fthenakis V (2013) Life cycle analysis in the construction sector: Guiding the optimization of conventional Italian buildings. Energy Build 64:73–89

Atmaca A, Atmaca N (2015) Life cycle energy (LCEA) and carbon dioxide emissions (LCCO2A) assessment of two residential buildings in Gaziantep, Turkey. Energy Build 102:417–431

Blengini GA, Di Carlo T (2010) The changing role of life cycle phases, subsystems and materials in the LCA of low energy buildings. Energy Build 42:869–880

Cellura M, Guarino F, Longo S, Mistretta M (2014) Energy life-cycle approach in Net zero energy buildings balance: Operation and embodied energy of an Italian case study. Energy Build 72:371–381

Cuéllar-Franca RM, Azapagic A (2012) Environmental impacts of the UK residential sector: life cycle assessment of houses. Build Environ 54:86–99
Dahlstrøm O, Sørnes K, Eriksen ST, Hertwich EG (2012) Life cycle assessment of a single- family residence built to either conventional- or passive house standard. Energy Build 54:470–479
EN 15978:2011. Sustainability of construction works. Assessment of environmental performance of buildings. Calculation method
Fouquet M, Levasseur A, Margni M, Lebert A, Lasvaux S, Souyri B... Woloszyn M (2015) Methodological challenges and developments in LCA of low energy buildings: application to biogenic carbon and global warming assessment. Build Environ 90:51–59
Gervásio H, Santos P, Martins R, Simões da Silva L (2014) A macro-component approach for the assessment of building sustainability in early stages of design. Build Environ 73:256–270
Gustavsson L, Joelsson A (2010) Life cycle primary energy analysis of residential buildings. Energy Build 42:210–220
Himpe E, Trappers L, Debacker W, Delghust M, Laverge J, Janssens A,...Van Holm M (2013) Life cycle energy analysis of a zero-energy house. Build Res Inform 41(4):435–449
Houlihan Wiberg A, Georges L, Dokka TH, Haase M, Time B, Lien AG,... Maltha M (2014) A net zero emission concept analysis of a single-family house. Energy Build 74:101–110
Iddon CR, Firth SK (2013) Embodied and operational energy for new-build housing: a case study of construction methods in the UK. Energy Build 67:479–488
Intelligent Energy Europe (2010) Enslic building: energy saving through promotion of life cycle assessment in buildings
Lewandowska A, Noskowiak A, Pajchrowski G (2013) Comparative life cycle assessment of passive and traditional residential buildings' use with a special focus on energy-related aspects. Energy Build 67:635–646
Monteiro H, Freire F (2012) Life-cycle assessment of a house with alternative exterior walls: comparison of three impact assessment methods. Energy Build 47:572–583
Motuziene V, Rogoža A, Lapinskiene V, Vilutiene T (2016) Construction solutions for energy efficient single-family house based on its life cycle multi-criteria analysis: a case study. J Clean Prod 112:532–541
Oregi X, Hernandez P, Gazulla C, Isasa M (2015) Integrating simplified and full life cycle approaches in decision making for building energy refurbishment: benefits and barriers. Buildings 5:354–380
Paleari M, Lavagna M, Campioli A (2016) The assessment of the relevance of building components and life phases for the environmental profile of nearly zero-energy buildings: life cycle assessment of a multifamily building in Italy. Int J Life Cycle Assess 21:1667–1690
Peuportier B, Thiers S, Guiavarch A (2013) Eco-design of buildings using thermal simulation and life cycle assessment. J Clean Prod 39:73–78
Praseeda KI, Reddy BVV, Mani M (2016) Embodied and operational energy of urban residential buildings in India. Energy Build 110:211–219
Proietti S, Sdringola P, Desideri U, Zepparelli F, Masciarelli F, Castellani F (2013) Life cycle assessment of a passive house in a seismic temperate zone. Energy Build 64:463–472
Rosselló-Batle B, Ribas C, Moià-Pol A, Martínez-Moll V (2015) An assessment of the relationship between embodied and thermal energy demands in dwellings in a Mediterranean climate. Energy Build 109:230–244
Rossi B, Marique AF, Reiter S (2012) Life-cycle assessment of residential buildings in three different European locations, case study. Build Environ 51:402–407
Stephan A, Stephan L (2016) Life cycle energy and cost analysis of embodied, operational and user-transport energy reduction measures for residential buildings. Appl Energy 161:445–464
Takano A, Pal SK, Kuittinen M, Alanne K (2015) Life cycle energy balance of residential buildings: a case study on hypothetical building models in Finland. Energy Build 105:154–164
Thiers S, Peuportier B (2012) Energy and environmental assessment of two high energy performance residential buildings. Build Environ 51:276–28
Zabalza I, Scarpellini S, Aranda A, Llera E, Janez A (2013) Use of LCA as a tool for building ecodesign. A case study of a low energy building in Spain. Energies 6:3901–3921

Chapter 5
Insights into AEC Firms to Survey Environment-Driven Change Management

Abstract To start delving into design and construction practices and ascertain how they are equipping and reorganizing in order to address environmental issues, a questionnaire survey was submitted to AEC firms. The aim is to provide a synopsis of the change management, focusing the attention on the transformation made by environmental issues and highlighting the inner structure, organization, competences and tools, encompassing cross-sectional fields. The target audience is Architectural and Engineering firms but also Construction companies established both at national and international level, surveying big- and medium-size firms, rather than small-size firms since less involved into the transformation process. Responses are analysed as group to determine trends on the change management of construction industry towards environmental sustainability, disclosing similarities and differences inherent design firms according to their size.

5.1 Questionnaire to Big- and Medium-Size AEC Firms

The global economic competition and the growing awareness of the importance of sustainable resource consumption are applying new pressure and demanding even more changes across AEC firms, both in terms of tangible and intangible resources (see Chap. 1). Architects, engineers, contractors and all the experts involved in the design and construction processes are exploiting the advancement in technology and in digital tools to improve sustainability, redefining step by step their workability and the organizational and management structure. Nowadays, more than ever, practice is in the paradoxical position of being invested in the production of real assets, working with tools of abstract representation, such as digital models and software simulations. The issue is that real world does not grow together with the virtual one, but rather remains behind, requiring long reaction times to comply with technological innovations. At the same time, AEC firms need to adapt to change, if they do not want to be exceeded by competitors (Lines et al. 2015).

A. Dalla Valle, *Change Management Towards Life Cycle AE(C) Practice*,
PoliMI SpringerBriefs, https://doi.org/10.1007/978-3-030-69981-9_5

To start gain insight on AEC perspective, a questionnaire survey was developed to analyse current design and construction practices and to understand how they are equipping and reorganizing themselves in order to address environmental issues. The aim is to provide an overview of the change management inherent AEC firms, focusing the attention on the transformation made by environmental issues and highlighting the inner structure, organization, competences and tools.

The target audience is Architectural and Engineering firms but also Construction companies, ranked both at national (Edilizia e Territorio 2014) and international level (ENR 2015). It is thus restricted to big- and medium-size firms, excluding from the scope small-size firms. Indeed, with regards to environmental issues, the transformation process gets involved especially big and medium-size firms rather than small-size firms, where the inner changes are limited and less visible. The questionnaire, structured with open-ended questions, was lunched on November 2015 and submitted to forty-six AEC firms. After one year, nine firms completed the survey (response rate 20%), resulting a small sample but, given the diversity of firms, good enough to depict current practice and identify ongoing trends. Among respondents, the majority are architectural firms (45%) or integrated firms (33%), followed by engineering firms (11%) and construction companies (11%).

In particular, the survey refers to four medium-size firms: Renzo Piano Building Workshop in Genova (ranked at the time at the first place in Italy), Progetto CMR in Milano (eleventh place in Italy), PiuArch in Milano (sixteenth place in Italy) and Mario Cucinella Architects in Bologna (thirty-second place in Italy); and five big-size firms, including HDR in Chicago (ranked at the time at the twenty-first place in the world), Arup in Berlino (twenty-fourth place in the world), HOK in Washington (seventy-eight place in the world), Foster and Partners in London and Skanska in Finland and UK (not mentioned in the rakings). Depending on availability, the questionnaire was conducted via mail, call or meeting, specifically asking for the involvement of environmental experts. Indeed, also the expertise, background and role of respondents strongly affect the outcome of the study, preferring where possible the interview of multiple participants to reflect different point of views and improve the reliability of data.

By the questionnaire it is possible to address a broad range of emerging topics, outlining the main differences and similarities inherent design and construction firms according to their size. The expertise and competence in the field, the types of works, the structure and organization, the network of offices, employees, and supporting partners as well as the use of BIM tools and the information flow are only few examples. Moreover, it deals with environmental issues, pointing out the main drivers, goals, experts, if any consultants and the use of Life Cycle Assessment as supporting tool in the decision-making process.

5.2 Organizational Re-Structure Between In- and Out-Sourcing

AEC firms responses are analysed as group rather than as individuals, to determine trends on the change management of construction industry towards environmental sustainability. For this purpose, they are grouped according to their size, considering as medium-size firms the companies established at Italian level and as big-size firms the companies established at international level, finding among them common ground but also diversities (Fig. 5.1).

The majority of firms turn out to be in both cases integrated structure, combining for example architects, structural engineering and HVAC for achieving sustainable buildings. In this way, the integration among architectural and engineering competences appears as a prerequisite to face the increasingly complexity of construction sector and to get sustainable design that ensures the control of costs, time and quality of projects (Keeler and Vaidya 2016). However, big-size firms consist of about 200 offices and 10.000 employees, as opposed to medium-size firms that are composed of around 10 offices and 80 employees. From practitioners point of view, the sprawl of many branches all over the world is a pivotal to be in close contact with local culture and standards, differing significantly from country to country. The head office of the AEC firms is usually placed in North America or in Europe with branches located especially in the Middle East and in Asia, but also in South America, Oceania and Africa.

Regarding competencies, two are the key features disclosed by the survey: the multidisciplinarity of actors involved, since multiple players are essential to design buildings at all scales; and the interrelated dependencies between experts, since no one reaches the goal alone but being part of a team. For this reason, AEC firms embed several areas of expertise in order to meet the wide range of requirements. The main specializations for architectural discipline are: architects, interior designers, landscape designers, industrial designers, urban planner and lighting designers; while for engineering they involve: mechanical engineers, structural engineers, plumbing engineers, electrical engineers, transportation engineers, environmental engineers, water quality engineers. In addition to these traditional actors, some AEC firms

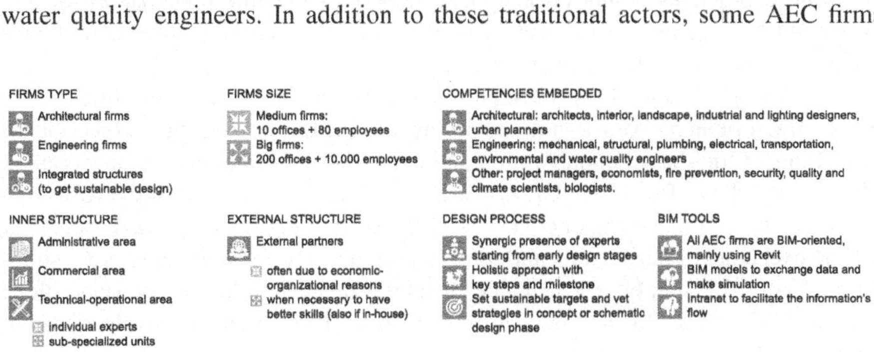

Fig. 5.1 Organizational structure of medium- and big-size AEC firms

integrate also other practitioners, such as project managers, economists, fire prevention specialists, technical and administrative specialists, acoustic specialists, environmental specialists, security specialists, quality specialists, climate scientists and biologists. Of course, not all the listed competences co-exist inside all design firms and their presence depends on the company size, the type of works and thus on regular customers. However, it is worth mentioning that all members are mutually responsible for the projects outcome and that each activity is fundamental to the overall output. Indeed, multidisciplinary approach and, to a lesser extent, also multicultural approach are becoming increasingly important, especially for integrated practice (Thiel et al. 2015).

The internal structure of the firms interviewed is always split in an administrative area, commercial area and technical-operational area. The latter is in turn divided in sub-specialized units to meet emerging issues, such as energy, environment, quality, safety and ICT, made up of individual experts within medium-size firms, while of specific teams within big-size firms. Although, not all specialized units with related experts are included in all design firms, their existence shows the growing interest in such matters and their strengthening over time to the point of requiring their establishment within the firm. Anyway, when for economic and/or organizational reasons there are no internal competences, AEC firms rely on external partners to meet specific design requirements. For medium-size firms this is recurrent, especially dealing with architectural and engineering firms, but also integrated practice. Whereas big-size firms, despite they embedded all expertise in-house and they are technically able to handle all project issues, sometimes they prefer to be supported by external strategic partners that know local requirements and have better skills in certain areas to enhance the projects.

All surveyed firms confirm the synergic action of the different experts starting from the beginning of the design process, outlining the uniqueness of each project characterized by different scopes and set-up. In this way, they prove the change management from the traditional linear process, where activities are performed sequentially according to project phases, to circular process, where all experts are involved starting from the outset, adding value and make improvements along decision-making. In addition, all practices include during process key steps and milestones, setting out the sustainable targets and strategies at the concept phase and checking their compliance with the progressive advancement of projects.

Concerning the digital transformation currently underway, all AEC firms claim to be BIM-oriented, even if not specifying the degree of maturity: object-based modelling, model-based collaboration and network-based integration (Succar et al. 2013). Regardless, Revit is the most widely used BIM software, in some cases still employed in conjunction with CAD software. BIM models are exploited especially for the exchange of design data and information, but also for running specific simulations, with the support for instance of energy analysis plug-ins or stand-alone software. Moreover, to facilitate the information flow among the wide range of actors involved, every AEC firms provide an intranet accessible to everyone from everywhere, to permit the team to work simultaneously from any office and place.

5.3 Mainstreaming Environmental Issues into Design Process

By surveying AEC practice, environmental issues are generally at the core of efforts, even if stressed in different ways according to firm size (Fig. 5.2). Inside big-size firms, they are constantly subject to review, in some cases just to meet regulatory requirements, but in the most virtuous cases because they share principles and action plans. Moreover, many companies have signed international and national policies, such as AIA 2030 Commitment in USA aimed at 70% energy reduction in all projects, regardless of project scope and requirements (Appelbaum 2015). By contrast, even if environmental issues are an integral part of design practice, medium-size firms are limited in the application of sustainable solutions since often not recognized as an added value by public and private clients. Indeed, there is a large gap in the awareness of environmental matters, with several conflicting opinions in the different countries. Nevertheless, the main environmental drivers are the growing awareness of the market, disclosed by both regulatory requirements and voluntary certifications, such as Green Building Rating System (Mattoni et al. 2018). In addition, the companies emphasise that nowadays a lot of private clients have adopted sustainable targets as part of their project standards, encouraging practitioners to seek new and better solutions for the built environment and to minimize building impacts.

Moreover, environmental issues affect the decision-making to a greater or lesser extent depending on the type of project, focusing over the past few years on energy and water and starting from 2016 on health and wellness. In particular, big-size firms usually handle all these aspects, whilst medium-size firms deal mostly with energy issues, since of interest to a wider group of clients and customers. The actors in charge are single specialist for medium-size firms and entire teams for big-size firms, trained in energy modelling, water, materials, health and wellness for delivering high performance projects. As individuals, they have the task to improve sustainable thinking within the team mindset, even if at the end it is expected from everybody an environmentally conscious approach during design. Nonetheless, both big- and medium-size firms sometime prefer to collaborate with third-party consultants, if strategic for the project together with all in-house expertise.

During design, every practice is accustomed to make environmental simulations of projects, choosing from the wide range of tools available on the market and able

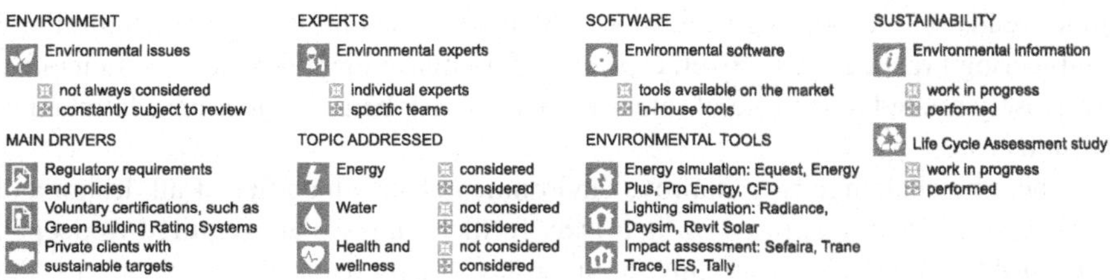

Fig. 5.2 Environmental concerns into medium- and big-size AEC firms

to meet every design demand, including: eQuest, Energy Plus, Pro Energy and CFD for energy simulation; Radiance, Daysim and Revit Solar for lighting simulation; Sefaira, Trane Trace and IES for impact assessment (Malkawi 2004). For this reason, interoperability is seen as a key factor for eliminating overlap, duplication of effort and to speed information flows (Augenbroe et al. 2004). Indeed, interoperability is intended as the ability of different information systems, devices and applications to access, exchange, integrate and cooperatively use data in a coordinated manner, to provide timely and seamless portability of information. At technical level, it implies the process of streamlining information exchange between two or more systems, while at organizational level, it refers to the ability for different stakeholders to work together towards common goals. During the design process, poor interoperability means to set limitations on the ability of the design team, denying their chance to be flexible with tools and often forcing them to narrow the range of software. In this context, while medium-size firms generally use the tools available on the market, big-size firms develop also in-house tools to suit specific project needs.

With regard to life cycle design, part of respondents uses to attach the environmental information of products to the BIM model, linking external LCA database rather than specific data. In particular, 44% of interviewees, notably big-size firms, perform LCA studies of projects, adopting the LCA tools available on the market, such as SimaPro for construction companies and Tally for integrated firms. Typically, the assessment primarily focuses on the technological components with greater impacts, such as concrete, steel and façade, and then working down to smaller elements. On the opposite, 56% of companies, represented by medium-size firms, do not develop LCA studies since not specifically requested by regulations and standards. Here, only in certain cases they estimate the environmental impacts of projects (expressed in tCO_2e/y), considering the energy consumption of the use stage. However, all AEC firms are aware that LCA becomes part of many Green Building Rating Systems (Dekkiche and Taileb 2016), calling for its implementation in practice, if they do not want to rely on external consultants focused on life cycle services.

Results shows how the environmental-driven change management brings out new pressure at organizational and business level, claiming the allocation of ever-increasing resources, both in terms of expertise and tools. The decision to structure internal firm hierarchies or to have external equipment depends on the cost that firms are determined to undertake in order to acquire knowledge, services or products. Note that due to the small sample population, the questionnaire survey allows to determine trends only according to firm size, flatten the peculiarity of architectural and engineering practices. Moreover, especially for environmental issues, the outcomes are closely related to the role of interviewees and risk being project-based, linked to virtuous cases.

Since the challenge is to improve environmental sustainability of all design and construction firms regardless of their size, further analyses are required to deepen current AEC practices, examining the interplay of experts, tools, requirements and targets as well as the information flow among the actors involved in the process. To this end, technology has great potential that, if skilfully exploited, can support

AEC firms in spreading environmental issues in a large number of projects and thus to reduce the environmental impact of construction sector. The problem is to understand where, when and how it makes sense to take action, since little and accurate solutions can lead to remarkable results.

References

Appelbaum A (2015) AIA 2030 Commitment, 2014 Progress Report. The American Institute of Architects

Arup (2016) https://www.arup.com

Augenbroe G, De Wilde P, Moon HJ, Malkawi A (2004) An interoperability workbench for design analysis integration. Energy Build 36(8):737–748

Dekkiche H, Taileb A (2016) The importance of integrating LCA into the LEED rating system. Procedia Eng 145:844–851

Edilizia e Territorio (2014) Top 100 società di progettazione. https://www.ediliziaeterritorio.ilsole 24ore.com

ENR (2015) Top 150 global design firms. https://www.enr.com

Foster and Partners (2016) https://www.fosterandpartners.com

HDR (2016) https://www.hdrinc.com

HOK (2016) https://www.hok.com

Keeler M, Vaidya P (2016) Fundamentals of integrated design for sustainable building. Wiley

Lines BC, Sullivan KT, Smithwick JB, Mischung J (2015) Overcoming resistance to change in engineering and construction: change management factors for owner organizations. Int J Project Manage 33:1170–1179

Malkawi AM (2004) Developments in environmental performance simulation. Autom Constr 13(4):437–445

Mario Cucinella Architects (2016) https://www.mcarchitects.it

Mattoni B, Guattari C, Evangelisti L, Bisegna F, Gori P, Asdrubali F (2018) Critical review and methodological approach to evaluate the differences among international green building rating tools. Renew Sustain Energy Rev 82:950–960

PiuArch (2016) https://piuarch.it

Progetto CMR (2016) https://www.progettocmr.com

Renzo Piano Building Workshop (2016) https://www.rpbw.com

Skanska (2016) https://www.skanska.com

Succar B, Sher W, Williams A (2013) An integrated approach to BIM competency assessment, acquisition and application. Autom Constr 35:174–189

Thiel S, Fiedler S, Loh J (2015) Multidisciplinary design in the age of data. In New challenges for data design: 353–375. Springer: London

Chapter 6
Assessment Framework Proposal to Steer Building Process Towards Life Cycle Perspective

Abstract Life cycle approaches are gaining even more attention in building sector, representing for the involved parties the key challenge for the coming years, asking practitioners to change their mindset and way of practice. To support AEC firms in life cycle design and operations, an assessment framework is developed based on current practice, with the aim to implement Life Cycle Thinking in building process, according to different process phases and empowering different actors. The focus is on the life cycle information required, the actors engaged to gather data and the tools and sources used to provide data. In this way, if explained from life cycle perspective, it helps in the data collection required to perform the inventory phase of life cycle studies (LCA and LCC). By contrast, reversing it from building process perspective, it supports the integration of LCT in practice, encouraging life cycle design and operations and the establishment of one-single life cycle information record for steering decision-making within a BIM-oriented working environment.

6.1 The Challenge to Match LCT and Building Process

The exploration of the current state of AEC firms, both by literature review (see Chaps. 1 and 2) and questionnaire survey (see Chap. 5), gives evidence of the change management of construction industry towards environmental sustainability. Indeed, while until a short time ago environmental targets were seen as constraints, today they are even more considered as a way to improve performance and increase competitiveness (Štefaňák 2011). In this context, life cycle approaches are gaining even more attention in building sector, representing for the involved parties a turning point for supporting sustainable practice, promoting environmentally-friendly innovation and business models (Antink et al. 2014) and improving resource efficiency with environmental, social and economic benefits. However, due to the complexities of building system, the wide range of requirements and the plurality of actors, the integration of life cycle approach in building process poses significant challenges, demanding within practice a shift both in thinking and in process.

The shift in thinking consists in treating buildings as unique systems, where each individual part affects and is in relationships with the others and the surrounding

A. Dalla Valle, *Change Management Towards Life Cycle AE(C) Practice*,
PoliMI SpringerBriefs, https://doi.org/10.1007/978-3-030-69981-9_6

environment, considering the entire life cycle. This imply not only a change in the mind-set from stuff (i.e. products and technologies) to purposeful systems- and life cycle- thinking, but also in the way of participating the building process. Indeed, in the age of specialization, one person cannot address all building aspects, different competences must be involved, bringing their specific know-how and interacting each other for optimizing looking at the whole and considering the entire life cycle. The shift in process demands thus a holistic process where everybody integrates their work rather than operate in isolated silos: the design team itself need to function as a system (Boecker et al. 2009). In this way, all design members have to understand how the decisions undertaken by each affect the decision made by all others, with the aim to jointly achieve sustainable and high-performance buildings.

As results, building sector demands a new process that encourages design teams and construction professionals to strengthen the two main tendencies in action. On one hand, the understanding of buildings in a systematic way. On the other, the interaction with a much higher level of communication and collaboration. For this purpose, an assessment framework is developed with the aim to integrate Life Cycle Thinking (LCT) in design and construction practice for orienting and streamlining building process in line with sustainable and environmental targets. To facilitate its implementation in AEC practice, it is envisioned within a BIM-oriented working environment, exploiting the potentialities of the tool currently adopted by practitioners to face the hard tasks distinctive for the construction sector (see Chap. 2). As outlined by BIM terms, "Building" concerns the physical characteristics of the model and stresses its capability to virtually recreate the facility considering the project-based tangible features. "Information" refers to the intangible characteristics of the model and stresses its capability to organize the set of facility data in a meaningful and actionable manner. Lastly, "Modelling" regards the act of shaping, forming, presenting and scoping the facility and stresses its capability to enable multiple stakeholders to collaboratively design, construct and operate the facility (Succar and Kassem 2015). BIM is therefore conceived as a database that embedded, display and calculates graphical/tangible and non-graphical/intangible information that, if properly arranged based on the proposed framework, provides a basis for decisions during the whole project life cycle.

The framework results from a matrix that combines life cycle perspective with building process. In particular, to put into effect LCT, that represents a general mind-set, Life Cycle Assessment (LCA) is taken as reference frame, providing an added value since depicts a standardized methodology at international level. In this way, the matrix of the framework is established, on the vertical axis, by the different stages of life cycle from cradle to grave and, on the horizontal axis, by the different phases of building process. The LCA stages follows the typical classification prescribed by standards (EN 15978:2011; EPD PCR UN CPC 531:2014): product stage (A1–A3), construction stage (A4–A5), use stage (B1–B7), end of life stage (C1–C4), benefits and loads beyond the system boundary (D). Each stage is detailed in the leading environmental topics in question: materials, energy, water, transport, waste and emissions, in turn specified according to LCA standards in the set of information to be considered. For this reason, "life cycle information" is used to refer not only to the complete list but also to clusters and individual information for remarking their relevance from a life cycle perspective. Instead, the building process phases are based on the harmonization of the different recognized partitioning (AIA 2014; RIBA 2013; UNEP 2014) in five main phases: concept phase, design phase, construction phase, in use phase and end of life phase. Note that despite the similarity of terms, life cycle stages not coincide with building process phases, since, for example, the design phase should take into consideration all life cycle stages. To avoid terminological misunderstandings, the term "stage" refers to life cycle methodology, while the term "phase" refers to building process.

Starting from the basic matrix, the framework interrelates building process with life cycle approach setting out: (i) the life cycle information required; (ii) the actors engaged to gather data; and (iii) the tools and sources used to provide data. The identification of these assets has been made through three main development steps. The first step, called "correlation", sets up for each life cycle stages and process phases the related information, actors and tools by means of hypothesis, providing the general concept of the assessment framework. The second step, called "identification", fills out the basic matrix with the information, actors and tools engaged in real building process, providing the testing of the assessment framework by analysing a sample of project case studies. The third step, called "redefinition", adjusts the suggested information, actors and tools based on current practice, providing the assessment framework proposal. In this way, the assessment framework represents a theoretical frame but practice-based. It recommends the most virtuous life cycle-oriented practice feasible nowadays, but its application depends case by case on how deeply life cycle perspective is or must be integrated in building process.

6.2 Stepwise Data Collection for Improving Life Cycle Studies

To face the complexity of the systems and to handle the large amount of information and data required to integrate LCT in building process, the assessment framework helps in the data collection required to perform the inventory of life cycle studies. Indeed, if explained from life cycle perspective, it shows for each life cycle information: the building process stage that deal with that type of data; the actors responsible for retrieving that information; and the sources and tools where to find information. In this way, if properly filled, it allows not only to arrange all life cycle information and relative data in a well-structured and comprehensive way, but also to understand the practitioners to collaborate with in relation to the building process phase at issue.

The proposed framework kicks off focusing on quantitative life cycle information as input data directly demanded by AEC firms for developing life cycle studies: Life Cycle Assessment (LCA) to determine environmental impacts (EN 15978:2011) and Life Cycle Costing (LCC) to determine economic impacts (ISO 15686–5:2017) along the whole building life cycle. The inventory phase of such studies represents not only the most challenging task, due to the large amount of data needed, but demands two type of information: quantitative information combined as appropriate with environmental/economic information. The first are directly tied to design and construction practice and therefore to bear in mind during building process, while le second refers to specific database usually embedded into the life cycle tools to be subsequently implemented. Note that if for LCA studies, the framework provides guidance for the data collection of the whole set of quantitative life cycle information, for LCC studies, it involves only building-related quantitative data input, excluding for instance the fees and taxes on goods and services. However, the consistency of both type of information significantly affects the level of completeness of the study and their quality affects the reliability and accuracy of the results. Information is expected to reach a greater level of detail as the process advances.

To provide an overview, the product stage (Fig. 6.1) requires the assessment of the whole set of materials, products and components used within the project for both construction materials and building systems. This information is available starting from the early process phases, being gradually designed and defined respectively by architects and engineers. In the concept phase, when design solutions are only sketched out, their quantification refers to preliminary design data, to be later detailed in the design phase with the connected bills of materials. During the construction phase, the actor in charge becomes the general contractor by means of bid data. Construction materials and building systems are subsequently keep in operation by the facility manager, tracing the related activities in the maintenance plan. Finally, during the demolition and deconstruction process the responsible actor is again the general contractor that, based on building data, sets the end of life of materials and systems to landfill or recovery, ensuring in the latter case their second life.

In accordance with LCA standards, a wide range of information are entailed for the construction stage (Fig. 6.2). They concern the transport from factory to building site and the connected materials missed or damaged during transportation that must be re-manufactured, whereas for the installation process they encompass almost all environmental topics. Materials for the elements missed or damaged during construction and for the requested ancillary materials. Energy for the consumption of equipment for ground work and landscaping, temporary works, on-site production, installation of products as well as of systems for storage of products, provision of HVAC and waste management process. Water for the usage for on-site production of product and for cooling of machinery and cleaning. Transport for both the handling of assets within site and the transport of waste from building site to landfill and, lastly, waste derived from construction process. Among this information, only few can be estimated during design by architects and engineers, considering for instance average distance for transport, literature data for energy and water used on-site production

					CONCEPT PHASE	DESIGN PHASE	CONSTRUCTION PHASE	IN USE PHASE	END OF LIFE PHASE
PRODUCT stage	A1-A3	Raw material supply Transport Manufacturing	materials	materials/products/components used within project for construction materials	*architects* design data [kg]	*architects and engineers* bill of materials [kg]	*general contractor* bid data [kg] [item]	*facility manager* maintenance plan [kg] [item]	*general contractor* building data [kg] [item]
				materials/products/components used within project for building systems (including energy generating units)	*engineers* design data [kg] [item]	*engineers* bill of materials [kg] [item]	*general contractor* bid data [kg] [item]	*facility manager* maintenance plan [kg] [item]	*general contractor* building data [kg] [item]

Fig. 6.1 Framework product stage

Stage	Module		Category	Item	CONCEPT PHASE	DESIGN PHASE	CONSTRUCTION PHASE	IN USE PHASE	END OF LIFE PHASE
CONSTRUCTION PROCESS stage	A4	Transport	materials	materials missed or damaged during transportation that must be re-manufactured			*sub-contractor* construction data [kg] [item]		
			transport	transport of materials/products/components from factory to building site		*architects and engineers* bill [t] average data [km]	*general contractor* bid data [t] [km]		
				transport of construction equipment to and from the site			*sub-contractor* bid data [item] [km]		
	A5	Construction installation process	materials	materials missed or damaged during construction and installation processes that must be re-manufactured			*sub-contractor* construction data [kg] [item]		
				materials ancillary for installation of products			*sub-contractor* bid data [kg] [item]		
			energy	energy used by equipment for ground works and landscaping			*sub-contractor* construction data [kWh]		
				energy used for storage of products			*sub-contractor* construction data [kWh]		
				energy used for temporary works (including the works off-site necessary for the construction process)			*sub-contractor* construction data [kWh]		
				energy used for on-site production of products			*sub-contractor* construction data [kWh]		
				energy used for the provision of HVAC systems during construction			*sub-contractor* construction data [kWh]		
				energy used for installation of products		*architects and enegineers* literature data [kWh]	*sub-contractor* construction data [kWh]		
				energy used for waste management process derived from construction phase			*sub-contractor* construction data [kWh]		
			water	water used for on-site production of products		*architects (wet/dry process)* literature / design data [l]	*general contractor* construction data [l]		
				water used for cooling of machinery or on-site cleaning			*sub-contractor* construction data [kWh]		
			transport	transport of assets (including materials, product, waste and equipment) within site			*sub-contractor* construction data [item] [km]		
				transport of waste from building site to landfill			*general contractor* bid data [t] [km]		
			waste	waste derived from construction process		*architects and enegineers* literature data [kg]	*general contractor* construction data [kg]		

Fig. 6.2 Framework construction process stage

and for waste generated. Conversely, all information can be tracked during construction process by general contractor or specific sub-contractor through bid data and construction data.

The use stage (Fig. 6.3) represents the extended phase of the process, establishing the service life of building and encompassing three macro areas. The first involves

Stage	Module		Category	Item	CONCEPT PHASE	DESIGN PHASE	CONSTRUCTION PHASE	IN USE PHASE	END OF LIFE PHASE
USE stage	B1	Use	emissions	emissions from the facade, roof, floor covering and other surfaces		*architects and enegineers* literature/design data [t voc]	*general contractor* bid data [t voc]	*sub-contractor* real data [t voc]	
	B2 B4	Maintenance Replacement	materials	materials/products/components used including ancillary products		*architects and enegineers* bill [kg] literature data [cycle]	*general contractor* bid [kg] maint. plan [cycle]	*facility manager* bid [kg] bid/real data [cycle]	
			energy	energy used for processes				*sub-contractor* real data [kWh]	
				energy used for waste management process derived from processes				*sub-contractor* real data [kWh]	
			water	water used for processes				*sub-contractor* real data [l]	
			transport	transport of materials for processes		*architects and engineers* average [km] literature [cycle]	*general contractor* bid [km] maint. plan [cycle]	*facility manager* bid / real data [km] [cycle]	
				transport of waste from site to landfill				*facility manager* bid / real data [km] [cycle]	
			waste	waste derived from processes				*facility manager* real data [kg]	
	B3 B5	Repair Refurbishment	materials	materials/products/components used including ancillary products				*facility manager* bid / real data [kg]	
			energy	energy used for processes				*sub-contractor* real data [kWh]	
				energy used for waste management process derived from processes				*sub-contractor* real data [kWh]	
			water	water used for processes				*sub-contractor* real data [l]	
			transport	transport of materials for processes				*facility manager* bid data [km]	
				transport of waste from site to landfill				*facility manager* bid data [km]	
			waste	waste derived from processes				*facility manager* real data [kg]	
	B6	Operational energy use	energy	energy used for heating system	*architects and engineers* software simulation [kWh]	*architects and engineers* software simulation [kWh]		*commissioning authority* real data [kWh]	
				energy used for air conditioning system	*architects and engineers* software simulation [kWh]	*architects and engineers* software simulation [kWh]		*commissioning authority* real data [kWh]	
				energy used for hot water supply system		*architects and engineers* software simulation [kWh]		*commissioning authority* real data [kWh]	
				energy used for ventilation system		*architects and engineers* software simulation [kWh]		*commissioning authority* real data [kWh]	
				energy used for lighting system		*architects and engineers* software simulation [kWh]		*commissioning authority* real data [kWh]	
				energy used for pumps, control and automation system		*architects and engineers* software simulation [kWh]		*commissioning authority* real data [kWh]	
				energy used for other building-integrated technical systems (e.g. lifts, escalators, safety and security installation and communication systems)		*architects and engineers*		*commissioning authority*	
				energy used for plug-in appliances		software simulation [kWh]		real data [kWh]	
				energy generated by renewable resources (e.g. photovoltaic cells, wind mills, solar thermal panels, ground or air source heat pumps)		*architects and engineers* software simulation [kWh]	*general contractor* bid data [kWh]	*commissioning authority* real data [kWh]	
				energy generated by non-renewable resources (e.g. gas boilers, combined heat power, fuels cells, heat pumps)		*architects and engineers* design data [kWh]	*general contractor* bid data [kWh]	*commissioning authority* real data [kWh]	
				energy used for cooking				*commissioning authority* real data [kWh]	
	B7	Operational water use	water	water used for drinking and cooking				*client* real data [l]	
				water used for sanitation system		*architects and engineers* software simulation [l]		*commissioning authority* real data [l]	
				water used for domestic hot water system		*architects and engineers* software simulation [l]		*commissioning authority* real data [l]	
				water used for irrigation system		*architects and engineers* software simulation [l]		*commissioning authority* real data [l]	
				water used for HVAC systems		*architects and engineers* software simulation [l]		*commissioning authority* real data [l]	
				water used for other building-integrated systems (e.g. fountains, swimming pools, sauna)		*architects and engineers* design data [l]		*commissioning authority* real data [l]	
				water used for appliances (e.g. dishwashers, washing machines)		*architects and engineers* design data [l]		*commissioning authority* real data [l]	

Fig. 6.3 Framework use stage

as a standalone topic the emissions from façade, roof covering and other surfaces, expressed in terms of VOC. The second area includes the whole set of operations for maintenance, repair, replacement and refurbishment processes. For each, the life cycle information concerns the required materials including ancillary products, the energy and water used for works, the transport of both materials to the building site and of waste to landfill as well as the waste derived. The third area regards building consumption for operational energy and water use, both broken down into the type of powered systems. All such information can be potentially gathered in the use phase, surveying for the first two areas facility managers and specialized sub-contractors and for the third area the commissioning authority. Nonetheless, given the specificity of data, the involved actors have to purposely track, monitor and record the information of their competence, installing when necessary sensors and meters. Anyway, some information can be anticipated during the design and construction phases, starting from the concept phase with energy simulations on heating and cooling. Furthermore, during design, additional information can be gathered about the emissions of selected surfaces, the materials and transport for maintenance and replacement processes, the operational use for both energy and water. In this case, the actors in charge are architects and engineers and the information source are, as appropriate, literature and design data. Finally, during construction phase, except for the predicted operational energy and water usage, the set of information can be updated based on bid data. The end of life phase is excluded from the scope.

The end of life stage (Fig. 6.4) claims the evaluation of the energy used for the de-construction and demolition processes, the transport of waste from building site to landfill, the waste derived, the energy used for waste processing, the transport of waste from treatment to landfill and the energy used for final disposal of materials. Among these, only few information can be considered starting from advanced design phase, esteeming the amount of waste and the related transport based on literature and average data. By contrast, all information can be potentially assessed during the end of life phase of building process, if intentionally tracked by general contractor and specialized sub-contractors.

Ultimately, the benefits and loads beyond the system boundary (Fig. 6.4), despite generally excluded from inventory, since pertaining to the second life or split between the first and the second life, embeds crucial information from a life cycle perspective. They cover the set of materials used for reuse, recycling and energy recovery and

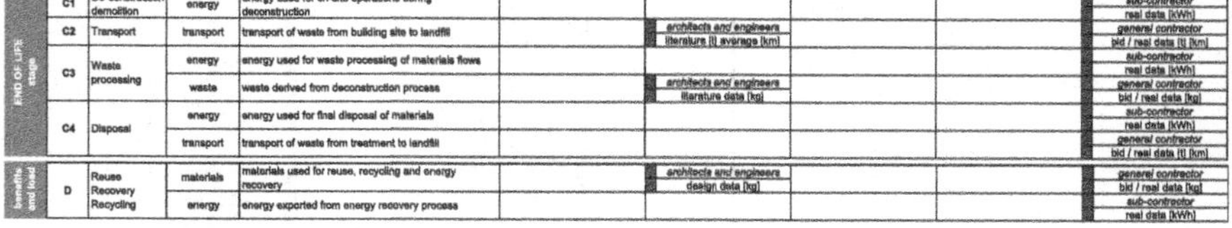

Fig. 6.4 Framework end of life stage and benefits and loads beyond the system boundary

the energy exported from energy recovery process. The amount of materials in question can be forecasted by architects and engineers during design, depending on the selected types of materials. Afterwards, they can be adjusted, updated and specified by the general contractor and sub-contractor during the end of life stage based on real data.

6.3 Life Cycle-Oriented Decision-Making for Process Optimization

Despite set up starting from life cycle stages, the framework can be reversed by building process, pointing out for each phase: the life cycle information to be accounted, the actors in charge of such information and the reference source and tools. In this way, it supports the implementation of life cycle practice, encouraging designers and practitioners in life cycle design and operations and orienting the decision-making to reduce building impacts and streamline the process.

Indeed, following the framework, the key actors of building process are notified from the outset of the information that form the basis of a life cycle approach and thus to bear in mind in the assigned duties. With the aim of pursuing a life cycle practice, they are invited to fill in the framework with the entrusted information, retrieving them on purpose from the stated sources and tools. Therefore, if properly completed, the assessment framework turns out to be a project-based and well-structured set of data of the facility along the whole life cycle, representing one-single record conceived in a BIM-oriented working environment. Collecting the quantitative life cycle information, practitioners are able to monitor, check, verify and encourage the process optimization towards sustainability. In fact, if life cycle data are lowered in value with the progressive advancement of the process, necessarily they cause low impacts. However, this statement is effective only when items of the same materials are considered during the building process (e.g. specific type of concrete), changing progressively the related quantities. By contrast, it falls when materials are replaced during the process (e.g. switching EPS with mineral wool). Here, the combination with environmental and/or economic data is demanded to make the different materials comparable in terms of impacts, developing thus simplified or complete LCA and LCC studies respectively. Exploiting BIM capabilities, the goal is to integrate the set of data starting from the early phase of the process, for providing a reliable basis for decision-making during the whole process. To this end, the framework is tailored to fit the peculiarities of building process phases and a progressive level of detail, empowering the responsible parties for the choices and activities taken in their expertise area and for the connected information flow.

To give a synopsis, the concept phase does not assume a key role in the collection of information but rather in setting targets to be achieved in the subsequent phases. Accordingly, the framework supports practitioners, including both designers and clients, in defining life cycle-oriented targets for the project, such as the percentage reduction of energy consumption, use of recycled materials and abatement of emissions. In addition, it encourages the long-term evaluation of the different concept design, like the choice to reuse existing structures or to opt for alternative solutions like expansions, renovations or new construction. The quantitative information available mainly refer to building structure and envelope, stressing design team in esteem the connected materials options in a life cycle way.

Towards life cycle perspective, the design phase should consider all life cycle stages, with the exception of repair and refurbishment, since they refer to activities that cannot be predicted in advance, integrating LCT into decision-making. For the product stage, as a common practice, they should choose the building components and systems, considering the relative amount of materials. For the construction stage, they should select the manufacturers not only in relation to the products and performance provided but also considering, for instance, the distance from the factory to the site. For the use stage, they should esteem the energy and water demand as well as the maintenance and replacement process of both materials and systems and the emission of finishes. Finally, for the end of life stage, they should account the materials diverted to landfill and the potential materials to be reused or recycled. Architects and engineers are responsible for the collection of the above-mentioned information, collaborating as appropriate with manufacturers. Concerning source and tools, in this phase a key role is played by bill of quantities and software simulations.

The construction phase, involving both the preparation of bid documents and the execution of works, must consider the information set in the design phase as thresholds for the decision-making process. In this way, general contractor and subcontractors are called for updating the information, with deepened data on construction materials and construction process. Here quantitative data turn out to be more accurate and reliable: the amount of materials and related data refer not to metric estimate but to tender documents and the information about construction and installation process refers not to literature but possibly to real data measured on site.

The use phase must monitor the building performance, focusing on the use stage for maintenance and facility process and for energy and water consumption. Indeed, during the operational phase, it is possible to compare, confirm or adjust the value derived from software simulations with the real consumption. Moreover, it is possible to check if the maintenance and replacement activities were confirmed as predicted in the previous phase, recording at the same time the information about repair and refurbishment operations. Here, the selection of the new building materials must be done with the same life cycle parameters adopted during the design phase and thus embracing from the production to the end of life stage of the products to be added. The actors in charge for gathering that type of data are facility managers and, if expected, the commissioning authority.

Finally, the end of life phase is intended to deepen, confirm or adjust the end of life and benefits beyond the system boundary information in relation to real data. As in the construction phase, the actors engaged are general contractors responsible for deconstruction, demolition, transportation, waste treatment and disposal or reuse, recycling and recovery process.

References

Antink R, Garrigan C, Bonetti M, Westaway R (2014) Greening the building supply Chain. UNEP sustainable building and climate initiative

Boecker J, Horst S, Keiter T, Lau A, Sheffer M, Toevs B (2009) The integrative design guide to green building. Wiley, New Jersey

Demkin JA (2014) The American institute of architects. The Architect's handbook of professional practice. New Jersey, Wiley

EN 15978:2011. Sustainability of construction works. Assessment of environmental performance of buildings. Calculation method

EPD PCR UN CPC 531:2014. Product Category Rules: Buildings

ISO 15686–5:2017. Buildings and constructed assets. Service life planning. Part 5: Life-cycle costing

RIBA (2013) RIBA plan of work 2013. https://www.architecture.com

Stefanák P (2011) Sustainability as a tool for increasing competitiveness. Studia Commercialia Bratislavensia 4(15):469–476

Succar B, Kassem M (2015) Macro-BIM adoption: conceptual structures. Autom Constr 57:64–79

Chapter 7
Shared Framing to Outline Environmental and Life Cycle Efforts into Building Design Projects

Abstract To bridge the gap on current environmental-driven change management, a shared framing is suggested for deepening building projects in terms of resources invested by design firms to meet environmental and life cycle issues. In particular, besides project general info, four areas of interest are dealt with: design team directory; environmental and life cycle studies; Green Building Rating System (GBRS) decision-making process; and assessment framework process. Depending on the scope, the shared framing is represented by linear graphs for individual analyses, while it is converted in circular graphs for collective analyses. The graphical representations provide added value: for external people, including clients, to have an overview of the projects/design firm aptitude for environmental and life cycle issues; for design firms to display in an effective way their environmentally-friendly way of practice; for practitioners to monitor the design process and enforce its optimization towards sustainability.

7.1 Ethnographic Approach to Fully Grasp AEC Practice

Since nowadays more or less every design firms claim to be environmentally friendly to take advantage for their business, the effort is to deepen how environmental issues are considered and faced in practice by AEC firms. Indeed while, on one hand, it is well known how sustainability and environmental topics are tackled at theoretical level by the scientific community and the wide range of possibilities offered by the tools now available on the market. On the other, it is mostly unknow how environmental issues are accomplished in practice within design and construction firms.

For looking deeply into their way of practice, the ethnographic approach is emphasised and applied to the construction industry, switching from the first model of interview, represented by the spread of questionnaire survey (see Chap. 5), to the second model of interview. The latter calls for a personal engagement within design practice for operating on-field analysis on a sample of project case studies. Ethnography in fact is an established qualitative research methodology, employed in a variety of contexts for finding answers to different issues, as commonly happens in sociology

A. Dalla Valle, *Change Management Towards Life Cycle AE(C) Practice*,
PoliMI SpringerBriefs, https://doi.org/10.1007/978-3-030-69981-9_7

to study the different way of life of certain populations. Concerning the construction industry, despite the inherent complexities, the field has remained under-theorized for a long time and ethnography is only recently emerging as part of the repertoire for understanding the building sector. To date there is a growing interest in and awareness of the utility of an ethnographic approach to the building sector, being defined taking into account how is practiced (Pink et al. 2013; Oswald et al. 2015).

Accordingly, the focus is on how AEC firms are equipping and reorganizing themselves in order to meet environmental and life cycle issues and their role within the decision-making process. To this end, a shared framing is developed with the aim to map the design process and point out all resources invested nowadays by the company to achieve environmental targets. In particular, they regard: the team of actors, including all disciplines and competences involved into design process; the set of technical design activities, including the development of environmental and life cycle studies and building sustainable certification; and the collection of data, including all information required both by experts and tools to work and design. For drawing it up, the traditional ethnographic method is demanded (Yaneva 2009), through extensive interviews with the actors involved in building process, such as architects, engineers and managers as well as external consultants and partners, and the review of project documents and tools.

Furthermore, in line with present trends that lead to consider buildings as small part of a larger networks, systems and environment, life cycle approach is stressed within the shared framing, as an ongoing and future challenge for AEC firms to take a broadening of perspective and to avoid shifting problems. When life cycle studies are applied to projects (e.g. Life Cycle Assessment and Life Cycle Costing), the goal is to figure out current practice to further improve their forcefulness and usefulness in the forthcoming design processes. By contrast, when due to a number of reasons life cycle studies are not developed, the goal is to figure out the life cycle information already available and their effective implementation in the process. Indeed, life cycle studies allow AEC firms to make aware decisions, gain long-term perspectives and define the more efficient way to meet sustainable requirements and decrease environmental, but also economic and eventually social, impacts.

The shared framing is narrowed to design process and it represents the methodology aimed at assessing the current design practices at the root of the proposed assessment framework (see Chap. 6). Notwithstanding it is intended for design practitioners to show their environmental and life cycle efforts to clients, to organize life cycle information in a systematic way, to verify the optimization or non-optimization of the process and to compare different design projects.

7.2 Linear Framing for Individual Development Over the Process

To support design firms in the acknowledgment and display of team efforts towards environmental sustainability and life cycle perspective, a shared framing is presented for examining design projects. The analysis is structured in general info and in four areas of interest: design team directory, environmental and life cycle studies, Green Building Rating System (GBRS) decision-making process and assessment framework process. Except for the preliminary information intended at providing a general overview of the project in question, all areas of interest are arranged in relation to design process phases, in order to disclose the leading topics and their impact on the decision-making.

For ensuring a user-friendly notice and a prompt understanding of the matter, each area of interest is represented graphically following a common layout, apart for the first stand-alone area. It consists on the phases of the design process that define thus the linear setting of the graphs. However, it is worth mentioning that the process linearity of the graphs does not mean and imply the linearity of the connected decision-making, since it is always expected to be iterative and recursive.

General info (Fig. 7.1) and the four areas of interest (Fig. 7.2) are explained in their general objectives, scope and set-ups, jointly constituting the shared framing with which depict the environmental and life cycle efforts of design projects at individual level.

The analysis kicks off with a brief description of the project in question and with the explication of its main features, including: client, location, building type, certification required, delivery method, BIM adoption, dimension, budget and schedule. The latter are represented in a cumulative graph, where the concentric arcs of circles stand for, starting from outside to inside: dimension (marked in green), budget for construction (in blue), budget for design (in red), schedule for construction (in blue) and schedule for design (in red). In this way, the graph allows to synthetically display the hallmarks of the project under investigation, offering a preliminary overview of its level of complexity.

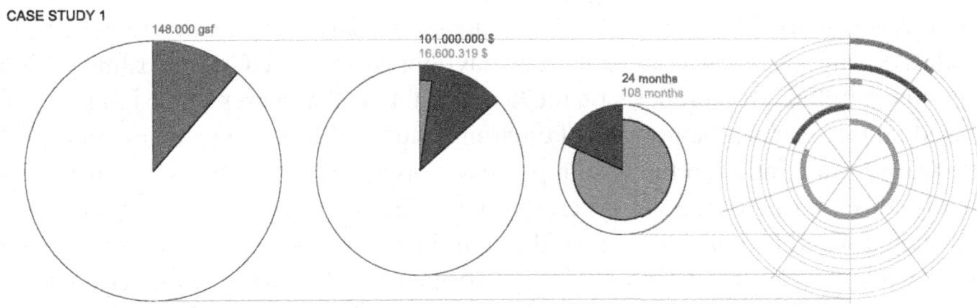

Fig. 7.1 Project general info underlying the shared framing

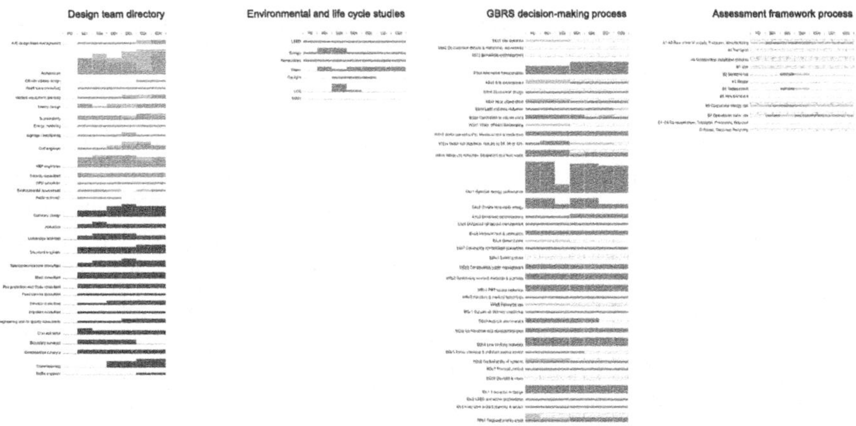

Fig. 7.2 Synopsis of the linear framing into the four interest areas

The first area of interest focuses on the design team directory required for the project at hand. The scope of the analysis is to understand not only the range of disciplines and people in charge, but also the complexity of the projects on the base of the multidisciplinarity demanded. The graph lists on the vertical axis the whole set of disciplines involved, while reports on the horizontal axis the phases of design process. Based on this matrix, the number of actors for each discipline is mapped distinguishing if internal (in red) or external (in blue) to the firm and highlighting, whenever possible, the subjects that remain equal to the previous phase and so permanent in the process (in dark red) or different from the previous phases and so new in the process (in light red). In this way, it is possible to figure out the specializations involved and, for each discipline, the change in the number of subjects with the connected crucial phases and eventually the transformation of the team of actors in relation to the design process phases.

The second area of interest focuses on the environmental and life cycle studies developed during the project. The scope is to understand which environmental and life cycle studies are integrated in the design process and therefore in which phase they affected the decision-making process. The graph lists on the vertical axis the environmental and life cycle studies, encompassing GBRS certification (e.g. LEED), studies on energy, renewables, water and daylight issues, Life Cycle Costing (LCC) and other studies, while reports on the horizontal axis the phases of design process. Based on this matrix, the number of environmental and life cycle studies (in green) are mapped in relation to the design process phases, except for GBRS certification that is constant to indicate the general filling of the scorecard without depicting the associated several studies. However, through the analysis it is possible to figure out in a synthetic way the design studies performed for the project to meet environmental and life cycle issues, their distribution within the process and eventually their confluence in specific phases.

The third area of interest focuses on Green Building Rating System certification and the connected decision-making process, if demanded for the project in question. The scope of the analysis is to give evidence to the set of sustainable topics designed and forecasted for the project, to achieve for instance LEED certification, and the variation of the credits claimed within the design process. The graph lists on the vertical axis the GBRS credits, while reports on the horizontal axis the phases of design process. Based on this matrix, the number of declared points is mapped, identifying from the scorecard the amount of points to be pursued for each credit (in dark green), to be further evaluated (in light green) or not reachable (identified only with a green profile). To this end, all prerequisites are excluded and when credits are split in sub-credits, they are grouped together representing the sum of the individual scores. In the credits that specify the amount of points considered as attainable by the project, the related possible points are not shown. Nonetheless, this is not valid in the case of grouped sub-credits, where the achievable points and the possible points can coexist since the scores are added together. Note that the list of credits is strictly related to the GBRS certification and related version in question and that the analysis allows to identify the accounted sustainable topics, their incidence as well as their changing along the process.

Finally, the fourth area of interest represents the core of the analysis and focuses on the life cycle information considered by the project at hand, applying the proposed assessment framework (see Chap. 6). The scope is to disclose the life cycle information accounted in practice during design and how they affect the decision-making process. The summary graph lists on the vertical axis the life cycle sub-stages, while reports on the horizontal axis the phases of design process. Based on this matrix, the explored environmental topics are mapped, embracing materials (M), energy (E), water (W), transport (T), waste (Ws) and emissions (Em) issues and revealing the intensity level with which they are taken into account during the process. The intensity levels are based on the number of times with which the studies on the topics are performed within the project process, marking them as high, medium or low (from dark to light green). Besides this, the tagging of the environmental topics means that the project team addresses them at the specific design phase, providing the associated information in relation to the life cycle stage at issue (see Chap. 6). What is important to stress is that environmental topics are closely related to the reference life cycle stages, for instance: the set of materials of the product stage are different from materials used for maintenance process, as well as the energy of construction installation process differs from the operational energy consumption. To remark the correlation of environmental topics and the reference life cycle stage, it is used the term "life cycle topics", referring thus to the whole, clusters or single environmental topics relevant for the stages of life cycle. Note that the graph does not disclose the complete list of life cycle topics demanded for each life cycle stage. In addition, the life cycle topics addressed specifically for GBRS certification, if pursued by the project, are shown separately from the ones related to environmental and life cycle studies, in order to underline the different drivers. As in the previous analysis, GBRS certification is always depicted as constant for all process phases, considering exclusively the life cycle topics stated in the last developed scorecard. In this way, the analysis focuses on design process, pointing out the assessed life cycle sub-stages with the related topics esteemed both for environmental and life cycle studies and for GBRS certification. Moreover, the intensity levels disclose in which process phases the life cycle topics influence more extensively the decision-making process.

As presented, the shared framing established for the project analysis at individual level allows a progressive deepening of the project design: from the general information and team directory, to the environmental and life cycle studies as well as GBRS certification, to the mapping of the explored life cycle information. However, it is worth mentioning that in practice it is crucial the phase at which the shared framing is adopted for the project analysis. If applied in the course of the project, it provides practitioners a user-friendly tool aimed, on one hand, at showing to clients the environmental and life cycle efforts undertaken during design, and, on the other, at monitoring the design process in order to improve its optimization towards sustainability. By contrast, if applied ex-post, it assists practitioners in mapping the design process of specific projects, focusing on environmental and life cycle issues but not affecting the decision-making process, thus failing short one of the main objectives. However, in both cases, the shared framing helps practitioners in pointing out the project aptitude for environmental and life cycle issues, by highlighting all resources invested by the firm to develop the specific building design project.

7.3 Circular Framing for Collective Development via Comparison

Since the analysis of individual projects is not exhaustive for outlining the design practice of a firm, the purpose is to broaden the project analysis from individual to collective level. To this end, the individual analyses on (i) design team directory, (ii) environmental and life cycle studies, (iii) GBRS decision-making process and (iv) assessment framework process are reworked to provide a common and comparable framing for all projects under investigation. The aim is to compare each other the design projects of a specific firm in order to figure out similarities and differences, identify trends and deepening design practice from an environmental and life cycle perspective.

Indeed, the shared framing proposed for the analysis at individual level mainly refers to the phases of design process which, changing from project to project, preclude the ground of comparing outcomes. To overcome this, in the analysis at collective level the shared framing is not based on design process phases, rather on the minimum and maximum number of assets, including both actors and environmental and life cycle studies, involved along the decision-making process. In particular, the minimum indicates the assets that are constant for each process phase, while the maximum discloses the assets variable in the process and that concern only specific phases. In this way, it provides an overview of the overall process, focusing on the variations of resources invested by the design firm.

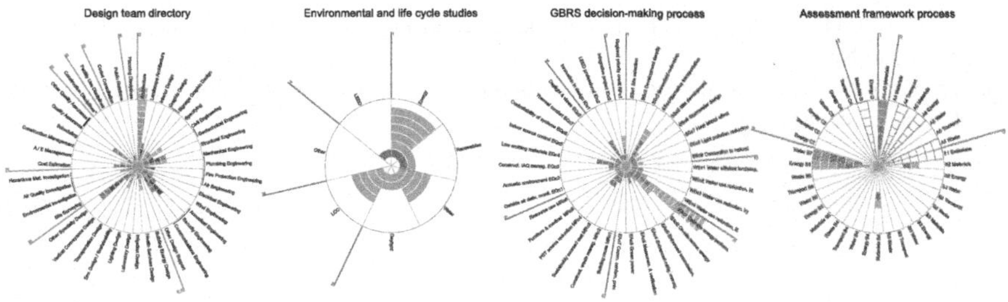

Fig. 7.3 Synopsis of the circular framing into the four interest areas

As for individual analysis, graphic representations are adopted to give a synopsis of the environmental and life cycle efforts spent during the design of projects and to promote a quick understanding of the matter. Contrary to the linear setting of the shared framing for individual analysis, the shared framing established for comparative analyses turn out to be circular (Fig. 7.3). Except for the general info graph which does not change, all graphs related to the four areas of interest get turned into a circular frame divided into wedges and in a series of concentric circles pertaining to the topics under study: design team directory (see Chap. 8), environmental and life cycle studies (see Chap. 9), Green Building Rating System decision-making process (see Chap. 10) and assessment framework process (see Chap. 11).

The application of the shared framing allows to compare in practice a sample of building design projects, revealing how design practice are organising and equipping in order to meet environmental and life cycle issues. By comparing a set of projects developed by the same company, the focus switch from project to design firm aptitude for environmental and life cycle issues. Unlike the individual project analysis primarily intended to the optimization of design process towards sustainability, the collective analysis aims to monitor the environmentally-friendly way of practice of a firm with a twofold purpose. On one hand, the mapping according to the complexity of the different projects of all resources invested by the firm to meet environmental and life cycle issues. On the other, in line with current trends, the life cycle perspective integration into design process. Indeed, the circular shapes of graphs facilitate not only the comparison of different design projects but also, to an extent, their assessment and level of maturity since they imply that the whole equals to complete (Klanten et al. 2008).

References

Klanten R, Bourquin N, Tissot T, Ehmann S (2008) Data flow: visualising information in graphic design. Gestalten, Berlin

Oswald D, Smith S, Sherratt F (2015) Accident investigation on a large construction project: an ethnographic case study. Proced Manuf 3:1788–1795

Pink S, Tutt D, Dainty A (2013) Ethnographic research in the construction industry. Routledge, London and New York

Yaneva A (2009) Made by the office for metropolitan architecture: an ethnography of design. Rotterdam, Lecturis

Chapter 8
Challenge of Managing the Design Team Multidisciplinarity in View of Environmental Goals

Abstract In term of resources invested by design firms to meet environmental and life cycle issues, the design team directory plays a key role, as evidenced by the first area of interest of the shared framing. The focus is on the competences and related number of actors involved for different projects, assessing the demanded level of multidisciplinarity and workflow management consistently with projects goals and features. By the comparison of a sample of building design projects, trends in current practice are outlined, stressing the wide range of specializations and their changes along the process, with the spread of energy and sustainable expertise. Some hotspots are further presented to improve the management of the engaged actors towards environmental and life cycle issues.

8.1 Key Competences of Design Team Directory

Starting from the shared framing (see Chap. 7), the first area of interest is represented by the design team directory, since crucial for highly competitive working environments, especially if with a view to sustainable development. For supporting practitioners in comparing the design team directories of a wide range of building design projects, a graphic representation is developed. The goal is to check the level of multidisciplinarity demanded, the completeness of the competences required and the amount of actors involved in the design, in respect of the expected project goals and building features. To this end, since in practice the competences are recorded anytime in different ways and with a multitude of terminologies, they were brought back to the OmniClass classification for disciplines (OmniClass 2012). OmniClass classification becomes thus the basic frame to bring into alignment and explain the set of disciplines and competences required to carry out the processes and procedures that occur during the life cycle of the different projects.

The graph (Fig. 8.1) consists of a circle split into forty-three wedges and in a series of concentric circles. The wedges reveal the complete set of competences demanded by the project at hand, grouped by the extended rays into the connected reference disciplines, while the concentric circles indicate the number of actors per process phase. Concerning competences, the hierarchical classification provided by

A. Dalla Valle, *Change Management Towards Life Cycle AE(C) Practice*,
PoliMI SpringerBriefs, https://doi.org/10.1007/978-3-030-69981-9_8

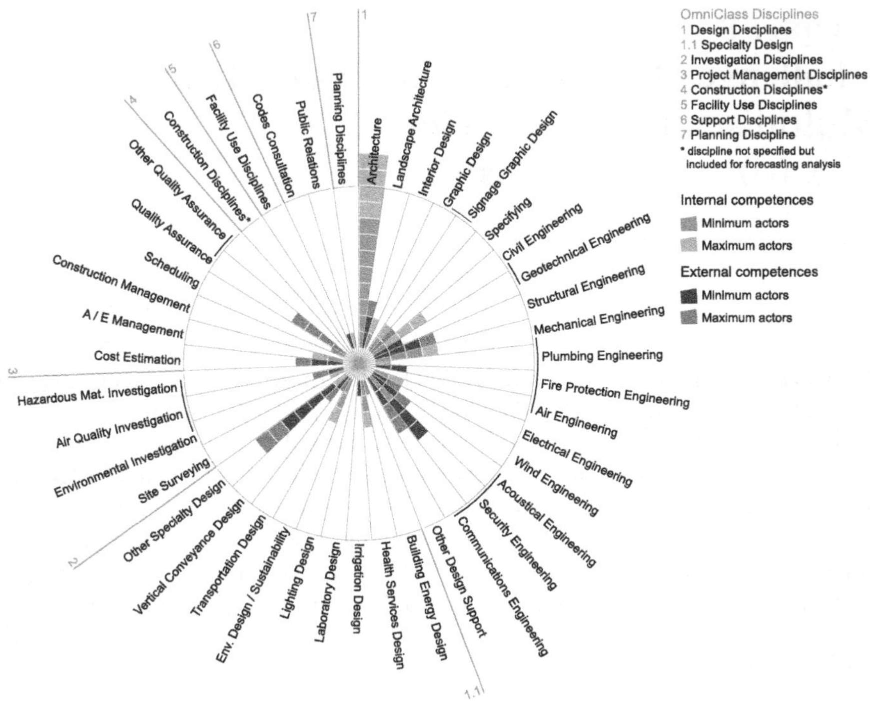

Fig. 8.1 Design team directory as first area of interest

OmniClass is here flattened to avoid disarrangement, linking only with an arch the competences that rely on a specific competence. In this way, for instance, plumbing, fire protection and air engineering are part of mechanical engineering; whereas acoustical, security and communication engineering are intended as specialty engineering. In addition, the competences engaged during the process are marked in the graph with different colours, to specify if entrusted to the design firm under study (in red) or to external companies (in blue). Each colour is in turn represented with different shades to denote the competences engaged for the whole design process (in dark red/blue), from the competences involved only in specific phase of the process (in light red/blue).

However, it is important to stress that some adjustments have been necessary to update OmniClass classification to the skill evolution required nowadays within building sector. Indeed, in line with current trends, the semantic meaning of certain competences has been extended over time, adjusting also the related terminology. This is the case of "building envelope design" converted into "building energy design", in order to include the so-called "energy modeller or analyst" as specified in most of project team directories. Replacing the term "envelope" with "energy", the aim is to expand the competence scope according to current needs, embracing not only envelope, but the overall relationship between architecture and building systems in terms of energy efficiency (Pacheco et al. 2012). Another transformation has been made for "environmental design" edited in "environmental design/sustainability" in order to broaden, based on present trends, environmental issues with economic and social issues.

Furthermore, with regard to design disciplines, the category "other" is added for including all those specializations that, despite not yet formalized and recognized by the classification, are required by current building projects. They involve for instance: off-site utilities design, cemetery design, food service consultant, helipad relocation and lab casework, emphasizing even more the increasingly specialization of design competences. Instead, concerning project management disciplines, "life cycle modeller" is included in "cost estimation" competence, giving evidence to the extension of the traditional cost estimation from building construction to the whole project life cycle. In addition, "commissioning" is embedded in "quality assurance" competence, as oversight and coordination of all planning, implementation and supervision of the project quality control plan, with the inspections and testing of construction practices (Djuric and Novakovic 2009). All these competences represent the principal set of expertise now called, in relation to project requirements, for the design and development of building projects.

8.2 The Widening of Competences to Face Project Complexities

By applying the graphical representation established for design team directory to different design projects, it is possible to depict in a uniform and comparable way the involved competences, on the basis of OmniClass classification. The purpose is twofold: on one hand, to identify current trends in the composition of project team and, on the other, to provide some hotspots for improving the management of the actors engaged towards environmental and life cycle issues.

Looking into a sample of building design projects (Fig. 8.2) developed by one of the top big-size firms established at international level and at the forefront of environmental issues (ENR 2014), the number of demanded competences appears closely related to the level of complexity of the projects. Indeed, from the comparison of graphs, P1-h and P4-h turn out to be distinguished by the broader spectrum of competences (about 60–70% of total), consistent with the high level of complexity determined especially by building type (hospital facility), federal requirements and LEED certification goals. All other projects required the involvement of a lower number of competences, ranging from thirteen to seventeen as appropriate.

Another point of attention is the correlation between the range of competences involved and the main features of the building to be designed. Usually, bigger is the project, wider is the range of competences engaged, except for P3-h because processed with Integrated Design Bid Built delivery methods. Building dimensions are also strictly linked to budget for construction, resulting in almost the same percentage, apart from P7-j where the percentage of construction cost is about half

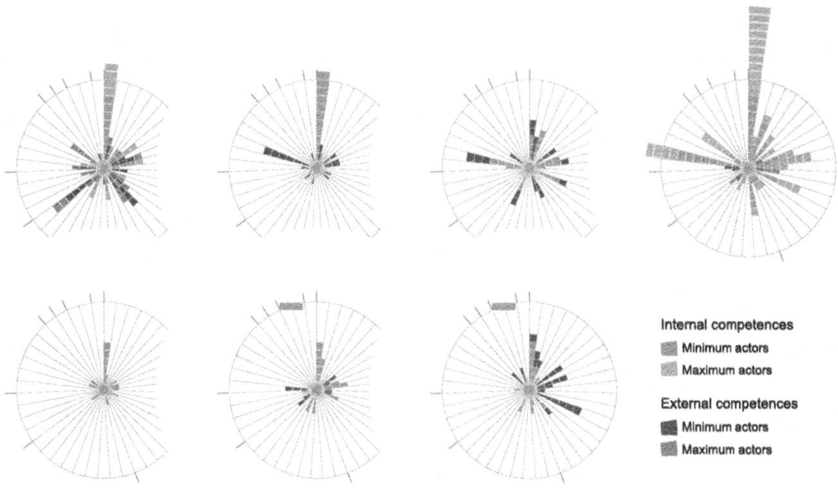

Fig. 8.2 Design team directory comparison of different building projects

of dimension. The budget for design is in direct proportion to both the range of competences involved and the level of complexity of the project, resulting except for P3-h infinitely less than construction cost. In addition, the time scheduled for design process is meaningful, since wider is the period planned, broader is the range of competences demanded. In this way, with the exception of few cases, the competences required for the different design projects are interconnected with the relative level of complexity, set by projects requirements as well as dimension, budget and schedule. The comparison of the design team directory allows thus to figure out how more complex and detailed the project are, wider is the range of competences and specializations. This means a further barrier from the managerial point of view, since more actors must be able to collaborate and work in synergy in order to identify the most appropriate design solution, also with reference to the environmental and life cycle goals.

Despite the design team directories are very miscellaneous and project-based, from the comparison it is possible to pick out some common aspects. With reference to OmniClass classification, two are the disciplines embraced in all projects team: the design disciplines, essential for the development of the overall building solutions, and the project management disciplines, crucial for the coordination of all involved disciplines and specializations. They are followed, as appropriate, by less widespread disciplines: the investigation disciplines, that survey and study the existing site conditions and the needed pre-design actions, and the support disciplines, that provide advice and assistance to business for programming and planning. The only disciplines not included during design are the facility use disciplines, responsible for the continued maintenance, operation and use of building and site after the construction, and the planning disciplines, in charge for the implementation of typically large-scale items. This is emblematic because it displays how nowadays the design decision-making process is not characterized by the disciplines intended to long-term vision (facility) and to systemic vision and extended context (planning). Hence the need to integrate these disciplines within the design team, as strategic priority especially for life cycle design.

In particular, concerning design disciplines, which encompass from architecture and engineering to specialty design, the design team of all projects (100%) involves the following competences: architecture, landscape, civil, structural, mechanical and electrical engineering. These competences represent therefore the primary set of expertise required during design process, regardless the type of building at hand and the level of complexity of the project. In addition, more than half of the projects (>50%) encompass further design disciplines, including interior, plumbing, fire protection engineering, building energy design, environmental design/sustainability and other specialty design, as well as some exponents of the project management disciplines, represented by cost estimation and quality assurance. Finally, the design team of less than half of the projects (<50%) requires the involvement of the following other competences: graphic design, signage design, specifying; geotechnical, air, wind, acoustic, security and communications engineering; other design support; health service, irrigation, laboratory, lighting, transportation and vertical conveyance design; site surveying; environmental, air quality and hazardous materials investigation; Architectural/Engineering (A/E) and construction management; scheduling; other quality assurance, codes consultation and public relations. The listed competences result thus less common within design team and their engagement in the different building projects is strictly related to specific requirements.

Nevertheless, among the wide and various range of competences included in the project directories, it is interesting to stress the spread of specialists focused on energy modelling and sustainability, disclosing how environmental issues play an increasingly important role in the decision-making process. Indeed, the analysed design team includes in most projects both competences, in few cases only one of them and just in P2-h none of them. Moreover, besides energy and sustainability specialists, life cycle modellers and commissioning prove how the emerging competences are pushing to a long-term vision of the building in compliance with economic aspects as well as environmental impacts and quality requirements.

As represented by the set of graphs, the design team directory is characterised by a variable number of actors for each competence, depending mainly on project requirement and ranging from a minimum of one to a maximum of nineteen people for each specialized expertise. Most competences are composed by few subjects, growing in specific area according to the distinctive traits of the project under study. However, by the comparison it comes to light the significant incidence of architectural competences, attesting how, despite the support of all other expertise, they continue to represent the cornerstone of design processes. This is noticeable in all building design projects, apart from P3-h where they are exceeded by A/E management competences, due to the specific integrated delivery method, and P7-j where electrical engineering emerges since relating to a jail facility.

In addition, it is worth mentioning that in some cases, P1-h and P4-h respectively, the number of actors per competence vary within the design process, as disclosed by the different shades of colour. Here, some disciplines are scheduled only for specific phases of the process, while others are required for all phases, affecting thus the whole decision-making process. Furthermore, some disciplines are "static", i.e. constant within the process, while others are "dynamic", i.e. characterized by a variable number of people within the process. Note that, even when for a discipline the number of actors is apparently constant, it is possible that the same people at issue change during the process, leading in some cases the change of the whole design team from one phase to another. Moreover, the complexity of the design team directory can increase exponentially because, even when different competences are part of the same design firm, often the actors are not in the same place but in offices located worldwide. By the comparison, it emerges thus how more complex the projects are, more likely the competences and/ or actors themselves vary within the design process, demanding a constantly check and management of the continuous changes.

8.3 Fostering Integrated Management of Actors and Know-How

As proved by the sample of building design projects, the design team is a complex system, intended to meet compelling design requirements and challenging sustainability goals. Indeed, depending on the level of complexity of the projects, it is characterized by a wide range of competences, by the variation of the number of actors, by the change of the same engaged subject, by their internally, externally or concurrent involvement and by their dislocation in offices placed worldwide. All these factors, carrying out continuous transformations in the organizational set-ups, further increase the complexity of the design process and the related information flow, especially if towards environmental sustainability.

To this end, the involvement of most competences in all process phases is pivotal, revealing how nowadays even more actors, specializations and competences are demanded simultaneously starting from the early stage of the process. It no longer exists the conventional linear process denoted by the progressive sequence of design activities: the design process demands even more an integrated, interdisciplinary, nonlinear and holistic approach, where everybody integrates their work rather than design within isolated silos. In addition, this statement is particularly crucial if the design process aims to integrate a life cycle perspective within the decision-making, requiring an even more increased collaboration, communication and cooperation among the actors.

In this context, two are the main hotspots to take into account for improving in practice the management of the engaged actors towards environmental and life cycle issues (Fargnoli et al. 2014). Firstly, the improvement of managerial competences to handle in the best way the changing directory and to monitor actors assignments in order to not weigh on the effectiveness of the process. Secondly, the implementation of new tools and methodologies in order to promote and foster the collaboration, communication, cooperation and the information flow between the different parties (Lee and Jeong 2012), as key strategies towards life cycle design.

References

Djuric N, Novakovic V (2009) Review of possibilities and necessities for building lifetime commissioning. Renew Sustain Energy Rev 13(2):486–492

ENR (2014) Top green building sector AEC firms. https://www.enr.com

Fargnoli M, De Minicis M, Tronci M (2014) Design management for sustainability: an integrated approach for the development of sustainable products. J Eng Tech Manage 34:29–45

Lee J, Jeong Y (2012) User-centric knowledge representations based on ontology for AEC design collaboration. Comput Aided Des 44(8):735–748

OmniClass (2012) A strategy for classifying the built environment. Table 33-Disciplines

Pacheco R, Ordóñez J, Martínez G (2012) Energy efficient design of building: a review. Renew Sustain Energy Rev 16(6):3559–3573

Chapter 9
Widespread Integration of Environmental and Life Cycle Studies for Sustainable Building Design

Abstract The set of environmental and life cycle studies are crucial to test in a preliminary way the effort invested by design team towards sustainability. For this purpose, the second area of interest of the shared framing focuses on the topics more widespread in practice, encompassing: sustainable certification, energy, renewables, water, daylight, life cycle studies and their impact on the decision-making process. In this way, by the comparison of different projects, it emerges the heterogeneity and uniqueness of each design process, in which only energy issues and sustainable certification turn out to be stationary topics, followed by studies on renewables and Life Cycle Costing. Based on current practice, some hotspots are suggested according to the level of sustainability and the maturity of life cycle implementation to be achieved within the projects.

9.1 Emerging Design Studies for Sustainable Buildings

Since towards sustainable building design the development of environmental and life cycle studies is pivotal, they represent the second area of interest of the shared framing (see Chap. 7). It is underpinned by a graphical representation raised in support of practitioners for mapping the design studies of a broad range of projects. The goal is to point out the topics that more affect the design process, disclosing their presence, constancy or variation and so their impact on the decision-making, in order to determine trends in current design practice.

The graph (Fig. 9.1) consists of a circle divided into seven wedges and in a series of concentric circles. The wedges reveal the environmental and life cycle studies, while the concentric circles indicate the number of times in which the studies are developed during the process. In particular, the design studies in question are grouped into the following categories: environmental studies, focused on energy, renewable, water and daylight; life cycle studies, including Life Cycle Assessment (LCA) and Life Cycle Costing (LCC); other studies and Green Building Rating System (GBRS) certification. The studies performed during the design process are marked in the graph, distinguishing between the ones conducted in a continuous way in all design

A. Dalla Valle, *Change Management Towards Life Cycle AE(C) Practice*,
PoliMI SpringerBriefs, https://doi.org/10.1007/978-3-030-69981-9_9

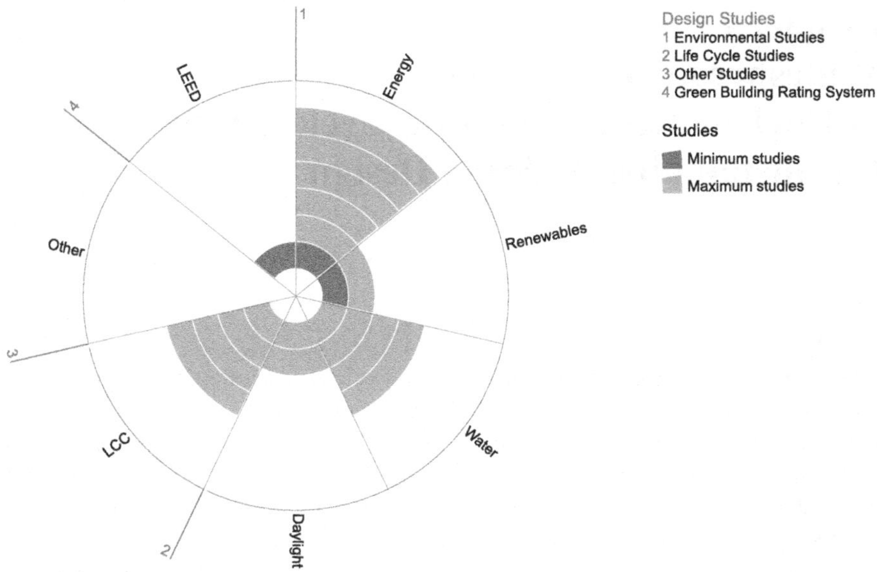

Fig. 9.1 Environmental and life cycle studies as second area of interest

phases (in dark green) and the ones developed occasionally, relying only on specific phases of the process (in light green).

Based on practice, other studies include for instance the assessment of building technologies for the achievement of labels, such as Energy Star roof (Sanchez et al. 2008), and the elaboration of Sustainable Return of Investment (SROI) studies, an enhanced form of LCC (Bichard 2015). To clarify, both LCC and SROI are especially required in the US context for federally owned and leased buildings (Code of Federal Regulations 1999), serving different purposes. LCC is a financial method of project evaluation in which all costs arising from a project (owning, operating, maintaining and ultimately disposing of it) are considered. In particular, potential energy or water conservation measures and renewable energy measures are assessed and verified in term of cost effectiveness. Indeed, the primary reason for an LCC is to demonstrate that the operational savings of a project are sufficient to justify its additional investment cost. By contrast, SROI incorporates also risk analysis and goes even further, providing a triple-bottom line view of the project economic results (Bohmholdt 2014). In this way, it includes economic impacts as well as monetizes (converts to monetary terms) all social and environmental impacts related to a given project. On that basis, the results of SROI studies follow the traditional LCC metrics, called Financial Return on Investment (FROI).

9.2 Studies Diversification in Line with Project Requirements

By surveying a sample of building design projects, it appears that the design process strictly depends on the overall requirements and goals to be fulfilled by each project. However, comparing with the support of the graphs the environmental and life cycle studies (Fig. 9.2), it is possible to assess how much the projects are environmentally/sustainability-oriented, to identify current trends and to provide shared hotspots for the enhancement of environmental and life cycle efforts.

During the decision-making process, the types of studies developed by the design team are closely connected to the level of complexity of the projects, rather than to the main features of the designed buildings. Indeed, they are generally not linked neither to the dimension nor to the cost. Only schedule affects directly the design process, since usually longer is the time for design, more studies are carried out. Indeed, the time planned for design is established in relation to the project requirements and it usually implies the level of complexity. In this way, environmental and life cycle studies result more integrated in federal projects, because of, especially if in conjunction with GBRS certification, they are targeted to a series of more restraining and challenging sustainable requirements. This is the case of P1-h, P3-h, P4-h and P6-l, characterized by both federal requirements and LEED certification and the development of almost all types of studies. They are followed by P5-l and P7-j with four types of different studies and finally by P2-h with only two types of studies, since represents an addition and renovation project.

From the comparison, it emerges how energy turns out to be the only stationary topic that appears in all design process, affecting more or less heavily the decision-making depending on the phases that deal with it. Thereafter, the environmental

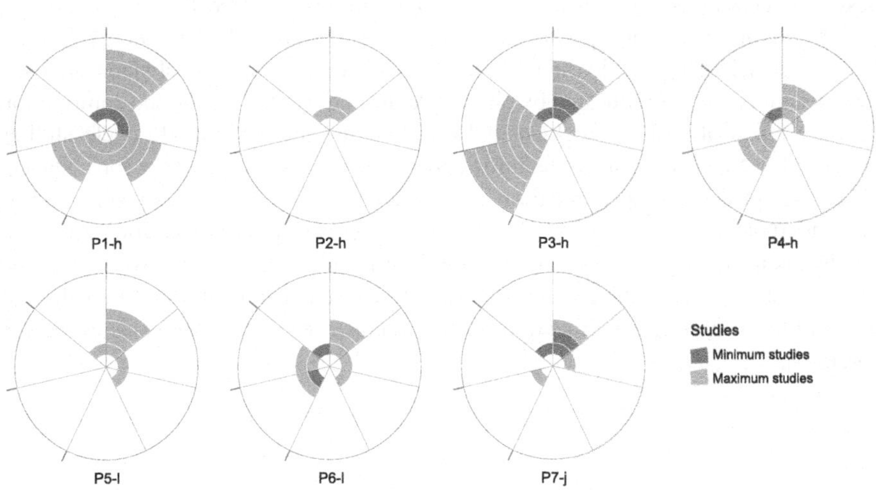

Fig. 9.2 Environmental and life cycle studies comparison of different building projects

and life cycle studies more spread in current practice are, in order of diffusion: at the second-place renewables studies; at third-place LCC studies; at fourth-place water and other additionally studies; and lastly daylight studies. In this ranking, however, it is important to highlight that some environmental studies, such as on renewables, energy, water and daylight, are expressly requested by LEED certification and mandatory to comply with project requirements, like energy simulations and, to a lesser extent, renewable studies. Moreover, note that the life cycle studies currently accounted in design practice assess only the economic impacts derived from project solutions along the whole life cycle (LCC), leaving behind the generated environmental impacts (LCA). This trend is partly countered by the development of Sustainable Return of Investment studies, which disclose how sustainability in its three pillars (environmental, economic and social) is claimed, especially by clients, as a new evaluation metric to prove and show the achieved design results.

Concerning the number of times in which such studies are carried out during the process, the projects turn out to be highly heterogeneous, performing as appropriate each type of study from one to seven times at each phase of the process. For instance, energy simulations are rerun six times in P1-h, only two times in P2-h and five times in P3-h. Nonetheless, it is worth mentioning that the number of times relies not only on project requirements but also on the effectiveness of the design process in the achievement of the established goals. Indeed, if a project is particularly challenging, pursuing thus high sustainable targets, it is possible that the design team takes some time before to find the most appropriate solution, reprocessing several times the studies in question. To understand how they affect the decision-making process, the environmental and life cycle studies are further explored in relation to design process. Indeed, some studies concern only specific phases of the process, while others are developed continuously. In most projects (70%), at least one environmental and/or life cycle study is carried out in a continuous way, involving thus all design process phases. Nevertheless, the steady studies appear with different combinations, involving as appropriate: LEED-energy-renewables, LEED-energy, LEED-LCC or only LEED certification. Typically, constant at design level since imposed by regulations and/or requested by clients are the energy studies and the compilation in these cases of LEED checklist. All other environmental and life cycle studies are strictly related to projects features and requirements. In this way, for instance, all federal projects have to perform LCC studies during the process, verifying as demanded the cost effectiveness of the adopted energy efficiency solutions. Anyway, the decision-making processes are in overall terms characterized by two joined and recursive design tasks: the continuous evaluation of environmental-oriented alternative and the subsequent verification of the savings obtained by the selected design solutions.

9.3 Balancing Studies Specialization with Knowledge Sharing

The comparison of the environmental and life cycle studies of a sample of building design projects allows to figure out, on one hand, the heterogeneity and, on the other, the uniqueness of each design process. In fact, the process depends primarily on how much the project is environmentally/sustainability-oriented, on the design requirements and on the specific goals set by clients.

In this context, the suggested hotspots are in direct proportion to the level of sustainability and of life cycle implementation to be achieved within the project. Higher is the level, more the environmental requirements and sustainability goals have to be shared by the engaged design team (Ghomashchi 2012). For this purpose, it is important to spread and integrate as much as possible the environmental competences, via specialized actors as well as common objectives and a responsible commitment of all involved subjects. Note that this results very challenging in the project characterized by a high level of complexity, since the shared goals have to be disseminated to a wide range of competences and actors.

In addition, more the project is inclined to environmental sustainability, more the environmental issues become the cornerstone of the process, requiring specific studies (depth) but also dissemination (extension) during the design. To this end, it is important to constantly update the inputs data, clearly identifying both the information required and actors involved in order to foster the communication and information flow between the subjects. Finally, more the project is environmentally/sustainability-oriented, more it is needed, on one side, the evaluation of different alternative solutions and, on the other, the check of the obtained savings. The results of the different studies have to be constantly shared, for instance through sustainable work session, to allow all actors to be aware of the impact of their decisions and to collaborate towards a common goal.

References

Bichard E (2015) Developing an approach to sustainable return on investment. Royal Institution of Chartered Surveyors

Bohmholdt A (2014) Evaluating the triple bottom line using sustainable return on investment. Remediat J 24(4):53–64

Code of Federal Regulations (1999) 10 CFR 436-Federal energy management and planning programs. Subpart A-methodology and procedures for life cycle cost analyses

Ghomashchi V (2012) Building sustainability through collaborative planning. Int J Sustain Dev Plan 7(1):14–25

Sanchez MC, Brown RE, Webber C, Homan GK (2008) Savings estimates for the United States environmental protection agency's energy star voluntary product labeling program. Energy Policy 36(6):2098–2108

Chapter 10
Green Buildings Rating Systems as Driver for Specific Life Cycle-Oriented Data Within Decision Process

Abstract The achievement of Green Building Rating System (GBRS) certifications further increase the efforts of the project team towards environmental and life cycle issues. That is why the third area of interest of the shared framing focuses on GBRS decision-making process, showing the list of credits and the related number of points considered as achievable, to be further evaluated or not reachable, for assessing the project level of sustainability. Surveying current practice, it emerges how the credits that provide specific life cycle-oriented information always affect the decision-making process of both design and construction phases, revealing therefore the availability of the connected data. They refer to water reduction, energy performance, construction waste, recycled content, regional and low emitting materials. For improving process efficiency, a brief synopsis is provided of the responsible actors and key issues to be addresses during design for pursuing in particular LEED certification and towards sustainability.

10.1 Leading GBRS Credits for Project Sustainability

Due to the widespread dissemination of Green Building Rating Systems as certification to evaluate the sustainability of construction projects, GBRS decision-making process turns out to be the third area of interest set by the shared framing (see Chap. 7). Deepening the environmental and life cycle efforts to achieve sustainable certifications, a graphical representation is provided to allow practitioners comparing several projects based on GBRS credits. The goal is to provide an overview of the different level of sustainability, pointing out the emerging issues in relation to the required project features.

The graph (Fig. 10.1) consists of a circle split in a series of wedges and concentric circles. The wedges reveal the set of credits established by the specific GBRS in question and clustered according to the pertaining categories, while the concentric circles indicate the number of points declared as attainable within the design scorecards. Concerning for instance LEED certification, the listed categories are: Sustainable Sites (SS), Water Efficiency (WE), Energy and Atmosphere (EA), Materials and Resources (MR), Indoor Environmental Quality (EQ), Innovation in Design

A. Dalla Valle, *Change Management Towards Life Cycle AE(C) Practice*,
PoliMI SpringerBriefs, https://doi.org/10.1007/978-3-030-69981-9_10

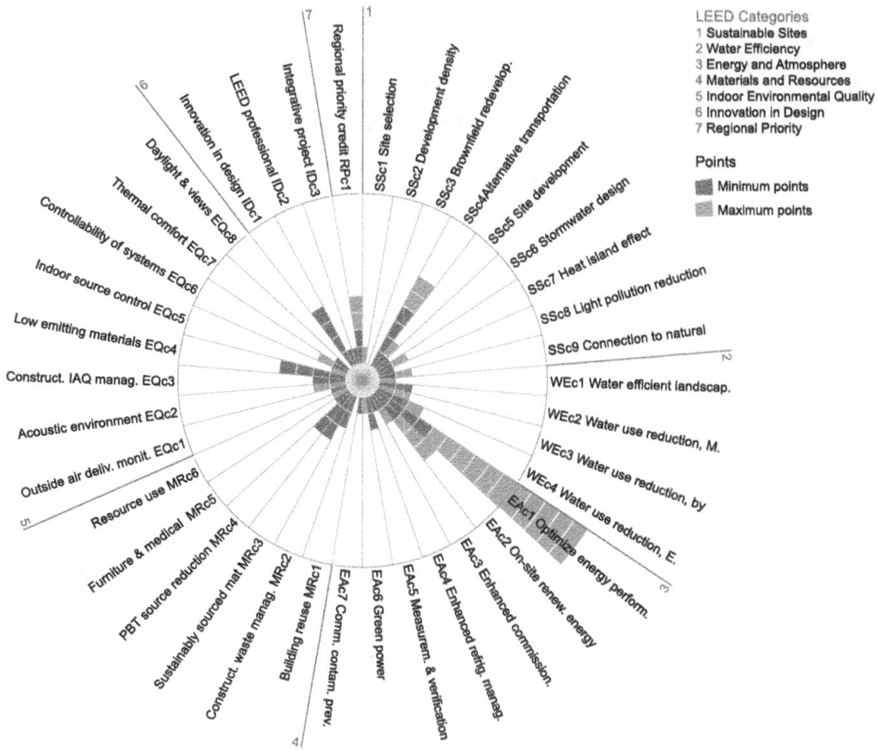

Fig. 10.1 GBRS decision-making process as third area of interest

(ID) and Regional Priority (RP). Based on this frame, the points declared as feasible during the design process are marked in the graph, discerning between the ones considered attainable in all phases (in dark green) and the ones accounted only in certain phases of the process (in light green). In this context, note that is possible to make direct comparisons only of projects assessed with the same type and version of GBRS.

10.2 In Silos Inclusion of LC Information for Project Certification

Among the broad spectrum of GBRS certifications now available on the market, LEED seems to be the more accomplished at global level (Cole and Valdebenito 2013), pushing the decision-making process towards environmental sustainability, as proved by the graphical representation. Indeed, its application to a sample of building design projects allows to identify the environmental and life cycle-oriented topics more affirmed in practice, recommending therefore the design team, also according to the assigned scores, to focus on certain aspects rather than others in the course of design (Suzer 2015).

From the projects comparison (Fig. 10.2), LEED rating system is always pursued along the process since expressly required by clients, except for P2-h where the design team carried out a LEED feasibility study to assess on a voluntary basis the project opportunity to achieve the certification. However, since the projects rely on different types and versions of LEED, it is possible to directly compare only: P1-h, P2-h and P4-h with reference to LEED for Healthcare 2009 (USGBC 2014); P3-h, P5-l and P6-l with reference to LEED for New Construction v2.2 (USGBC 2005); and solely P7-j with reference to LEED for New Construction 2009 (USGBC 2015). Anyway, since the majority of credits, despite arranged in different ways, is shared among the different types and versions of LEED, all LEED decision-making processes are analysed together in order to figure out trends and provide some hotspots to improve the sustainability of future projects.

Regardless of the main project features, like dimension, budget and schedule, all designed buildings are distinguished by a wide range of credits considered as attainable along the process. GBRS certifications in fact are generally followed to reach high sustainability goals and thus only if goods results are expected. That is evidenced by the sample of design projects, where they all earn high scores: Gold certification is achieved for P1-h, P3-h, P6-l and P7-j; Silver certification for P4-h and Platinum certification for P5-l. Conversely, low score is obtained for P2-h, in which LEED certification was not required by client and thus the design team was not encouraged to adjust from the outset the decision-making to comply with LEED criteria.

However, looking into the set of credits declared as feasible with the relative points, the resulting configurations are representative of the peculiar key features of the projects, in compliance with inherent characteristics and design goals. Some graphs

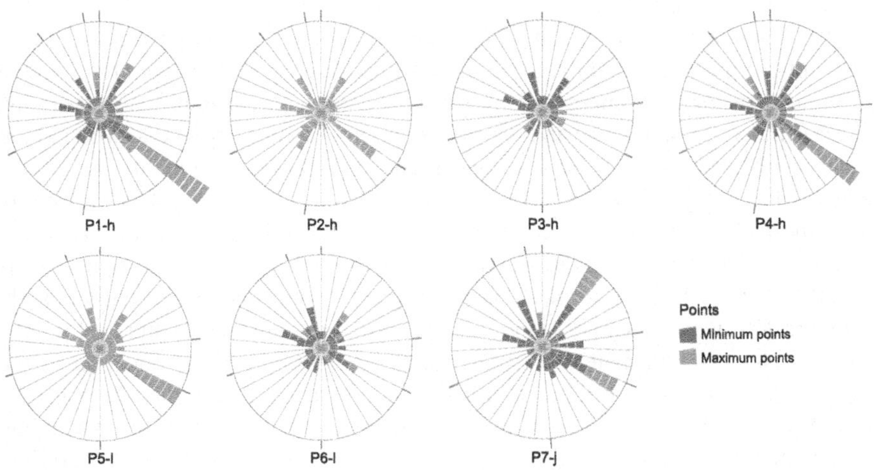

Fig. 10.2 GBRS decision-making process comparison of different building projects

show significant peaks pertaining to specific credits or categories. In particular, peaks are attributable to energy performance optimization, because involving the credits with the highest associated number of points and therefore taken by practitioners as primary goals during the decision-making process. Only in P7-j, the points of energy performance are exceeded by alternative transportation ones, just because the representation adds together all points achieved within the connected sub-credits.

Narrowing the analysis to the credits that provide worthy information from a life cycle perspective, LEED for Healthcare 2009 is selected as reference for nomenclature, covering anyway the credits shared by all types and versions of LEED. Indeed, jointly considering the LEED scorecards of the different projects, it is possible to note the recurrence of some credits, allowing to figure out the life cycle topics that more affect the decision-making process in current practice. Always accounted within the projects (100%) are building water use reduction (WEc3), optimise energy performance (EAc1), construction waste (MRc2) and low emitting materials (EQc4). Moreover, all healthcare facilities certified with LEED HC 2009 achieve the specific credits on equipment water use reduction (WEc4) and sustainable sources materials (MRc3). If the first credit is peculiar solely of LEED for Healthcare, the second is included also in LEED for New Construction, although split in separate credits according to the topics in question. Indeed, both versions of LEED NC divide it into the following five credits: materials reuse, recycled content, regional materials, certified wood and rapidly renewable materials. Here, the credits on recycled content, regional materials and certified wood are accomplished by all projects, whereas rapidly renewable materials by half of them and materials reuse by only one project. In addition, other topics commonly taken into account to improve buildings sustainability are water efficient landscaping (WEc1), spread in 85% of projects, and on-site renewable energy (EAc2), ascertainable in 70% of projects. By contrast, the projects never deal with building reuse (MRc1), since they mainly constitute new buildings without chance to reuse existing structures, in any event considered not advantageous for the few renovation projects.

In this context, nevertheless, it is important to stress that the design team is responsible only for part of the appointed credits, while the construction team represented by the contractor is in charge of the remaining ones. In particular, all credits on Water Efficiency and Energy and Atmosphere relies on design process, whereas the credits on Materials and Resources and Indoor Environmental Quality depend on construction process (Lohiya 2013). Energy and water issues therefore constantly affect the decision-making process, forcing the design team to develop several studies in order to select the design solution that achieve the highest number of points. In this way, they are regularly performed during design but, while energy rises significantly in the graphs, water does not stand out due to the low number of attached points. The same happens for materials issues, which impact more or less consistently the different projects. However, since in this case contractor is the main responsible, nowadays materials selection affects only indirectly the design decision-making, often inducing the design team to check a priori the related credits, without carry out any verifications.

It is important to stress that the presented LEED decision-making refers exclusively to design process, omitting the subsequent processing occurred during construction (Kubba 2009). In this regard, despite some credits can be revoked during LEED review process, the projects comparison serves to identify the issues that more affect the design decision-making, as evidenced by the variation of points. Indeed, some credits display consistent improvements during design, getting higher scores from one phase to another, as result of their impact on the decision-making process.

10.3 Systemic Approach Over Building Life Cycle

The GBRS process comparison, here restricted to LEED certification, focuses on the credits more representative from a life cycle point of view, neglecting the ones that provide further specifications on compositional and functional aspects. The analysis has a twofold objective. Firstly, the understanding of the life cycle information and data to be taken into account during design and to be made available along the process. Secondly, the comprehension of the underlying mechanisms of the certification process, with the connected credits phases and actors in charge for the fulfilment of credit requirements and related information flow. In this way, assuming the building design projects as reference practice, it is possible to provide some hotspots to improve the efficiency of the design process towards sustainability.

Especially in projects where clients aim at achieving a high level of certification, it is strategic to identify from the early phases of the process the targets set by GBRS credits, adopting them as shared goals. Moreover, concerning the credits officially committed to contractor, the design team has to anticipate as much as possible the selected alternative solutions, in order to put forward and thus optimize the process. Instead, regarding the projects not intended for pursuing any certification, it is at least advantageous for the design team to gather from GBRS credits the specifications to be followed during the process, considering them as reference values for the decision-making.

For this purpose, a brief synopsis is provided of the key issues to be addresses during the process (Shan and Hwang 2018) for pursuing in particular LEED certification and towards sustainability. Such issues are explained, in line with current trends, taking a life cycle perspective to show in a systematic way the information potentially available to develop the most varied design studies. Furthermore, since all LEED types and versions specify for each credit the same reference phase and practitioners, the responsible actors are entrusted in tackling as best practice the recommended issues during the decision-making process, considering for deepening the related LEED targets.

Concerning product stage (A1–A3), the conservation of natural resources is encouraged by promoting the use of existing structure, envelope and elements, calling the contractor for evaluating the quantity of preserved materials. Moreover, a series of specifications is provided for the choices of both construction materials and building systems (Arroyo et al. 2016). Architect and interior designers are tasked with the use of salvaged/refurbished/reused, rapidly renewable and wood-based materials and products, to support contractor in making decisions and evaluating the percentage of building materials that meet the specific criteria. In addition, addressing healthcare facilities, the selection of sound and noise insulation materials is assigned to acoustic engineers, while the verification of the chemicals content of furniture and medical furnishings products to interiors, owner and specialists. Still at the design phase, further criteria affect the building materials selection process, focusing for instance on the solar reflectance index of roof and paving materials. By shifting to engineering solutions, the team is invited to perform special duties and, according to the established criteria, to integrate specific building systems into projects. Civil engineers are involved for the management of stormwater runoff; whereas electrical engineers for the verification of interior and exterior illuminance value, for the chemical content and mercury use of lighting products and devices, and for the implementation of controls lighting systems of all multi-occupant spaces. Finally, mechanical engineers deal with refrigerant management of HVAC systems, the indoor air quality of mechanically and naturally ventilated spaces and the controls system of individual thermal comfort.

Regarding construction process stage (A4–A5), both transport and waste issues are focus of attention. Contractor is motivated to select a share of building materials and products that are extracted, harvested, recovered as well as manufactured within the region. Moreover, contractor is in charge of tracking the amount of construction waste for detailing the materials to be diverted from disposal, such as: the recyclable/recovered resources to be redirected to manufacturing process and the reusable materials to be brought back to appropriate site.

Moving on the use stage, surface emissions (B1), energy consumption (B6) and water consumption (B7) must be at the core of design decision-making process (Greer et al. 2019). Indeed, the adoption of low emitting materials is promoted for interior adhesives and sealants, flooring, composite wood and agrifiber elements, committing the products selection to contractor but also architects, interiors and specialists. In healthcare facilities the checking compliance with VOC limits is further demanded to interior designers for furniture and medical furnishing products. Furthermore, mechanical engineers have to verify the contaminant releases to air from products of combustion and to plan permanent monitoring systems for CO_2 concentrations to sustain occupant comfort and well-being. About energy, due to the key role played, the energy analyst in coordination with mechanical engineers continuously run energy simulations to define, check and compare the energy profile of the proposed building with the baseline model. This is in conjunction with the energy produced by renewable systems and the use of grid-source renewable energy technologies, pushing electrical engineers, energy analyst and sustainable champion to install photovoltaic and solar thermal panels and to evaluate the purchase of green power. In addition, implementation of commissioning activities is endorsed to verify the energy-related systems performance, as well as the installation of meters to measure building energy consumption and occupant thermal comfort over time. Another topic of interest is the operational water use, repeatedly assessed by plumbing engineers considering building fixtures, fittings, appliances and process equipment, for verifying the reduction of both the potable water demand and generation of wastewater. The installation of meters to provide the ongoing accountability and optimization of building water consumption performance over time is also suggested. Furthermore, civil, landscape and irrigation specialists are requested to eliminate the use of potable water for irrigation, by using only captured rainwater and recycled wastewater and installing landscaping that not require permanent irrigation systems. The use of potable water is further discouraged in healthcare facilities, accounting building system equipment, cooling towers, food waste systems and all medical equipment.

Finally, the end of life stage (C1–C4) is overlooked, whereas concerning benefits and loads beyond the system boundary (D), the focus of contractor, supported by architects and interiors, is on the selection of construction products that incorporate recycled content, involving both post-consumer and pre-consumer materials.

In this way, especially if pursuing LEED certification, the project team is impelled to deserve special attention on the above-mentioned issues, aligning the decision-making towards appropriate solutions in order to comply with credits requirements and thresholds. However, note that for improving environmental sustainability, it is more appropriate to establish the percentage based on mass and not, as required by LEED credits, based on costs. Moreover, looking into credits from a life cycle perspective, it emerges that they tackle only with specific life cycle topics, without showing an overall and systemic vision of the designed building. This is introduced only starting from later version of LEED (USGBC 2016), demanding the development of Life Cycle Assessment studies and the comparison between the proposed building and the baseline model.

In broad terms, it is important to highlight that the project information worthy from a life cycle perspective are fragmented in the most varied GBRS credits and assigned to a wide range of competences. Only the synergy between the actors, the efficient exchange of information and the integration of the proposed solutions makes it feasible to achieve good scores and to provide high-performance sustainable buildings.

References

Arroyo P, Tommelein ID, Ballard G (2016) Selecting globally sustainable materials: a case study using choosing by advantages. J Constr Eng Manag 142(2):05015015

Cole RJ, Valdebenito JM (2013) The importation of building environmental certification systems: international usages of BREEAM and LEED. Build Res Inform 41(6):662–676

Greer F, Chittick J, Jackson E, Mack J, Shortlidge M, Grubert E (2019) Energy and water efficiency in LEED: How well are LEED points linked to climate outcomes? Energy Build 195:161–167

Kubba S (2009) LEED practices, certification, and accreditation handbook. Butterworth-Heinemann

Lohiya G (2013) Contractors and LEED credits: towards sustainable construction. In: ICSDEC 2012: developing the frontier of sustainable design, engineering, and construction, pp 885–893

Shan M, Hwang BG (2018) Green building rating systems: global reviews of practices and research efforts. Sustain Cities Soc 39:172–180

Suzer O (2015) A comparative review of environmental concern prioritization: LEED vs other major certification systems. J Environ Manage 154:266–283
U.S. Green Building Council (2005) LEED for New Construction & Major Renovations v2.2
U.S. Green Building Council (2014) LEED 2009 for Healthcare
U.S. Green Building Council (2015) LEED 2009 for New Construction & Major Renovation
U.S. Green Building Council (2016) LEED v4 Building Design + Construction

Chapter 11
Heterogeneity and Fragmentation of Life Cycle Approach in Current Building Design Projects

Abstract Since life cycle approach is still not fully established in design firms, the fourth area of interest of the shared framing is aimed at showing the life cycle information considered in practice along the process. The focus is on materials, energy, water, transport, waste and emissions of the different life cycle stages, marked according to their level of intensities in relation to design phases. By the comparison of a sample of building design projects, it emerges the partially coverage during design studies of the life cycle information, limited to few construction materials and building systems, energy consumption and, to a lesser extent, water consumption, seldomly enlarged to construction waste, use emissions and maintenance materials. In this context, further efforts are needed to improve life cycle design in practice, as suggested by the hotspots.

11.1 Core Life Cycle Topics Along Design Process

As life cycle approach is stressed as an ongoing and future challenge for AEC firms, the fourth area of interest of the shared framing (see Chap. 7) dealt with it, surveying its application in practice. For investigating in a systemic way the life cycle topics currently considered in building projects, a graphical representation is developed from the proposed assessment framework (see Chap. 6), correlating the design process of different building design projects with life cycle perspective. The comparative analysis has a double purpose. On one hand, it allows to figure out the life cycle topics now accounted by design team and their level of impact during the decision-making process. On the other, it discloses how life cycle perspective is actually integrated into design practice.

The graph (Fig. 11.1) consists of a circle split into thirty-nine wedges and in a series of concentric circles. The wedges reveal the specific life cycle topics (including materials, energy, water, transport, waste and emissions) referred to all life cycle stages, standardized into product stage, construction process stage, use stage, end of life stage and benefits and load beyond the system boundary (EN 15978:2011). Whereas the concentric circle indicates the different design process phases of the project at hand (AIA 2014). Based on this, the different shades of colour depict

A. Dalla Valle, *Change Management Towards Life Cycle AE(C) Practice*, PoliMI SpringerBriefs, https://doi.org/10.1007/978-3-030-69981-9_11

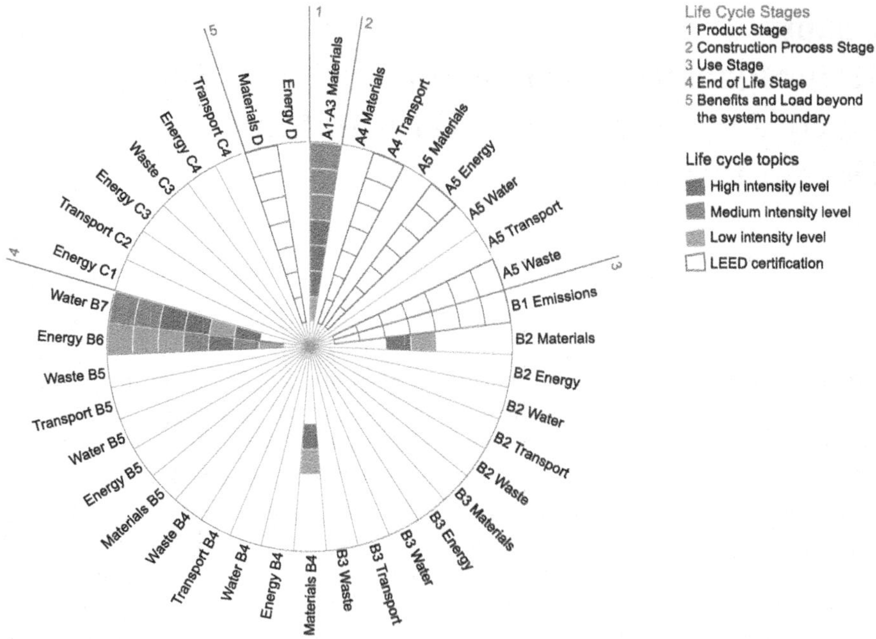

Fig. 11.1 Life cycle approach as fourth area of interest

the level of intensity with which life cycle topics are considered along the design process. In particular, three level of intensity are marked in the graph, resulting as high, medium and low intensity level in relation to the number of times in which studies on the topics are performed within the design process. Life cycle topics are thus mapped according to the environmental and life cycle studies developed during design as well as in line with Green Building Rating System (GBRS) certification process. When topics overlaid, priority is given to the life cycle topics addressed by environmental and life cycle studies, not depicting the ones associated with GBRS certification.

Note that graphs elaborated for different projects are not plainly comparable for two main reasons. Firstly, because the design process is partitioned in different way within the projects, ranging as appropriate from three to seven phases. Secondly, because the intensity levels of the life cycle topics are assessed by comparing the design studies referred to each individual project. Notwithstanding, the indirect comparison of the graphical representations is enough for the scope of the analysis. Moreover, it is worth mentioning that, among the four areas of interest of the shared framing (see Chap. 7), this is the only case in which design process stands out from the circular graphs.

11.2 Project Requirements Focused on Few LC Information

The decision-making processes of a sample of building design projects are explored in depth according to the graphical representation derived from the simplification of the proposed assessment framework (see Chap. 6). Focusing on the life cycle information considered along the process and displaying them in a synthetic way, the variety of design processes comes up from the graphs, showing how they are basically project-based and linked to the main building features and requirements. However, by the comparison (Fig. 11.2), it is possible to figure out what are the life cycle topics accounted in current practice and how they affect nowadays the decision-making process. This step turns out to be crucial to grasp and outline current design practice with reference to life cycle perspective as well as to identify the basic hotspots to enhance their impact in the future.

To date, the design process takes into consideration very few life cycle topics compared to the total number of topics that occur during the whole building life cycle, emphasizing how Life Cycle Thinking is only partially integrated into design. This state does not rely too much on the main project features, such as dimension, budget and schedule, but rather on requirements, resulting therefore as one of the main life cycle drivers. Indeed, the projects developed in compliance with federal mandates appears from a life cycle perspective more complete than the ones that meet standard requirements. All ordinary projects, represented by P2-h, P5-l and P7-j, consider during the decision-making the principal construction and systems materials for the product stage and the operational energy for the use stage, combined only in certain cases with operational water use and materials used for maintenance process. Instead, federal projects as P1-h, P3-h, P4-h and P6-l enrich the above-mentioned life cycle

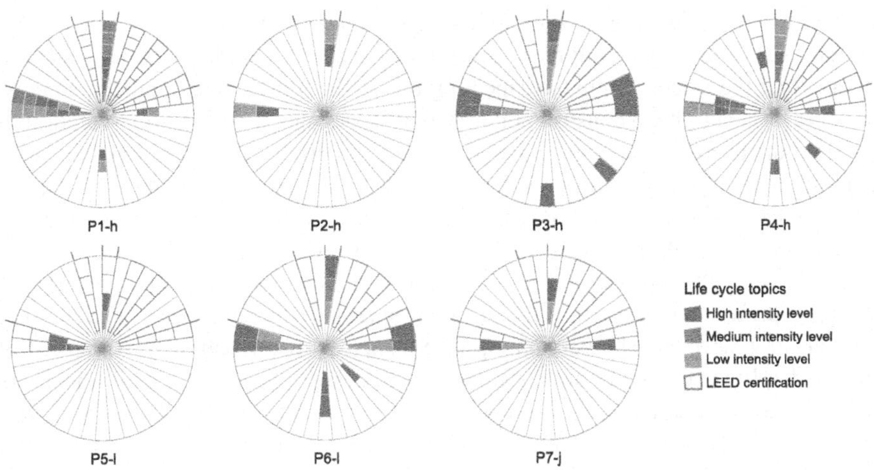

Fig. 11.2 Life cycle approach comparison of different building projects

topics including in all cases for the use stage the operational water and in some cases also emissions. Half of federal projects encompass also further life cycle stages and topics, for instance taking into account construction waste and recycled materials.

As advanced, the life cycle topics that affect the decision-making process depend strictly on the type of environmental and life cycle studies developed by the project team during design. Considering the entire sample of reference projects, the topics more widespread in current practice (100% of projects) are production materials and energy use, since energy simulations and renewables studies are now almost established (Baird 2018). They are followed by operational water and maintenance materials (70%), since included in water studies and Life Cycle Costing (LCC). In addition, dealing with LCC studies, replacement materials (55%) and repair materials (40%) come after in order of diffusion. Less common within the decision-making process are use emissions (30%), waste construction and recycled materials (15%), since embedded in niche studies, like Sustainable Return of Investment (SROI) and studies on technologies. Furthermore, it is worth mentioning that such emerging life cycle topics are always explored not in their full extent but focusing only on the main technological elements and/or specific components under investigation.

However, by adding the issues faced for the achievement of LEED certification, a more extensive spectrum of life cycle topics is included in the assessment. Indeed, some LEED credit documents provide specific life cycle information of the designed buildings (see Chap. 10) enhancing, on one side, the cited life cycle topics and, on the other, revealing further life cycle topics. Except for P2-h that not pursue any certification, all others (100%) deepen a share of both construction materials and building systems, to meet credits requirements and specifications. Energy for installation is indirectly considered by commissioning and construction management plant credits (Harris et al. 2020), while construction waste is always defined and clearly communicated by contractor. Concerning the use stage life cycle topics, emissions from surfaces are taken into account, selecting low emitting materials and systems, as well as building energy consumption, renewables production and water use for systems, landscape and equipment. All projects include in addition materials benefits beyond the system boundary, meeting the credits on recycled content. Finally, in some projects (85%) construction transport is accounted through the selection of local products, and maintenance materials are indirectly assessed through enhanced commissioning. In any event, since formally construction team (contractor) is responsible for credits on materials (Ferme et al. 2018), they do not necessarily affect the design decision-making process, unless bring forward by the design team.

To delve into the impact of life cycle topics on the decision-making process, the graphs represent along the rays the different phases of design process, unlike other circular graphs of the shared framing (see Chap. 7). If the full slice is marked, it means that life cycle topics are considered at all design phases; instead if it is partly coloured, it implies that they are considered only in specific phases. In this way, current projects turn out to be very heterogeneous and characterized from a life cycle perspective by incomplete and typically fragmented processes. However, it is interesting to stress that in the more complex and demanding projects, P1-h, P3-h, P4-h and P6-l respectively, most life cycle topics are taken into account starting from the early phase of the process and along the whole decision-making. In particular, this continuity is evidenced for operational energy and envelope materials and, to a lesser extent, for operational water, since they represent crucial issues during design. By contrast, the life cycle topics derived from LCC and SROI studies are usually isolated within the process, apart from P6-1 where they are more diffused. Nonetheless, it is significant to underline the phases of the process in which they occur. LCC studies are generally carried out at the core phases of the process to verify the cost effectiveness of a range of design options in order to select the most suitable solution (Goh and Sun 2016). Conversely, SROI studies are developed only at the end of the process, thus not affecting design choices and acting as a communication and validation tool for disclosing to clients the results achieved (Bohmholdt 2014).

Lastly, the intensity level with which the cited life cycle topics are taken into consideration during the decision-making process is very miscellaneous. In some cases, they involve mainly the initial phases, in others the central phases and in others again just the final phases, disclosing how they are project-based.

11.3 Integrating Full LC Information to Orient Decision-Making

The assessment framework process comparison allows to provide an overview of current practice focusing on the life cycle topics now considered in decision-making. Results show that actually life cycle perspective is only partially established, life cycle topics affect just to some extent design process and the life cycle information now available are scattered in the most varied studies and documents. In particular, they are restricted to few construction materials and building systems, energy consumption and, to a lesser extent, water consumption, enlarged by LCCA, SROI studies and LEED certification to construction waste, use emissions, maintenance materials. In this context, further efforts are needed to improve life cycle design in practice, as suggested by the following hotspots.

To provide a more complete overview and to avoid shifting problems, the design team is strongly encouraged to consider the widest range of life cycle topics and set the connected information, also based on literature data (see Chap. 4). For this purpose, since not all life cycle topics can be foreseen during design, the attention is called only to the information suitable for the phases of the process (see Chap. 6), gradually including the connected data to steer decision-making and to develop the most varied design studies. Such studies are defined according to project complexities and requirements and must be implemented starting from the early phases of the process to impact as much as possible design choices. Indeed, it is of limited outlook to perform studies at the end of the process because, not affecting decision-making, they do not support design in providing sustainable solutions.

In this context, the collection in one-single record of all project life cycle information as well as their progressive implementation is crucial to not slow down design process but rather to optimize it.

References

Baird G (2018) Energy performance buildings. CRC Press

Bohmholdt A (2014) Evaluating the triple bottom line using sustainable return on investment. Remediation J 24(4):53–64

The American Institute of Architects (2014) The Architect's handbook of professional practice. Wiley, New Jersey

EN 15978:2011. Sustainability of construction works. Assessment of environmental performance of buildings. Calculation method

Ferme L, Zuo J, Rameezdeen R (2018) Improving collaboration among stakeholders in green building projects: role of early contractor involvement. J Legal Aff Dispute Resol Eng Constr 10(4):04518020

Goh BH, Sun Y (2016) The development of life-cycle costing for buildings. Build Res Inform 44(3):319–333

Harris F, McCaffer R, Baldwin A, Edum-Fotwe F (2020) Modern construction management. Wiley

Chapter 12
Existing Information Flow to Develop Environmental and Life Cycle Studies of Current Practice

Abstract Staring from the premise that there is not a predetermined relationship between the resources of a firms and its capabilities, the purpose is to figure out how they are linked together, revealing the information flow in action towards environmental sustainability. To this end, the collaboration and communication required is deepened, in order to understand how current practice work nowadays to meet environmental and life cycle issues, also in relation to the level of complexity of the projects in question. Among the complete list of competences, the responsible actors are thus highlighted, disclosing the competences that need to collaborate each other to provide and to gather the information required for the different studies. Results evidence how, on one hand, only few actors are in charge of the environmental and life cycle studies and, on the other, that a widespread network is demanded, recommending sustainability and energy specialists to immediately start collaborating and working in synergy with the main supporting disciplines.

12.1 Cross-Competence Collaboration Towards Sustainability

In the context of environmental-driven change management faced by construction industry, the proposed shared framing (see Chap. 7) is intended to show the environmental and life cycle efforts of design firms, surveying in particular: the design team directory (see Chap. 8), the environmental and life cycle studies (see Chap. 9), the Green Building Rating System decision-making process (see Chap. 10) and the assessment framework process (see Chap. 11). However, it is solely by a conjunction of all these aspects that it is possible to provide an overview of current design practice. To this end, the focus is on the information flow (Baldwin et al. 1999), combining the set of involved actors, performed design studies and considered information to meet environmental and life cycle goals. The aim is to deepen the collaboration and communication required, in order to understand how current practice work nowadays to meet environmental and life cycle issues, also in relation to the level of complexity of the projects in question. To provide a synopsis, the organizational set

A. Dalla Valle, *Change Management Towards Life Cycle AE(C) Practice*,
PoliMI SpringerBriefs, https://doi.org/10.1007/978-3-030-69981-9_12

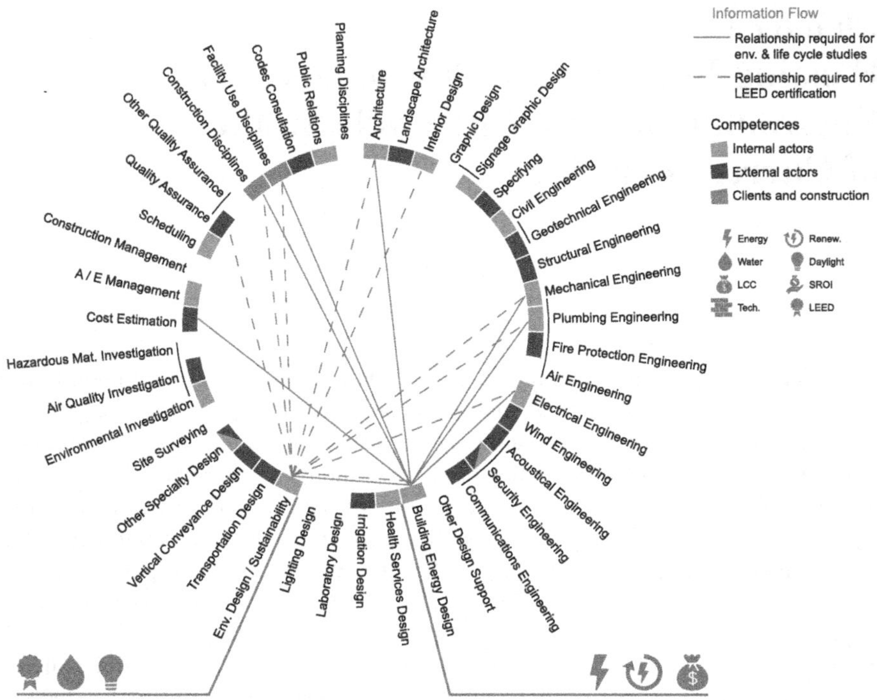

Fig. 12.1 Information flow of current design practice towards environmental sustainability

up occurred during design projects towards environmental sustainability is displayed and graphically synthetize, consistently with the ones developed for the specific areas of interest.

The graph (Fig. 12.1) consists of a ring split into forty-three wedges corresponding to the complete list of OmniClass competences (OmniClass 2012) engaged in the sample of building design projects (see Chap. 8). In this way, it depicts the design team directory, distinguishing the competences demanded internally to the design firm under investigation (in red) and assigned externally to third parties (in blue). The competences in charge of the environmental and life cycle studies are then underlined, pointing out for each the specific studies developed, including: energy, renewables, water, daylight, Life Cycle Costing (LCC), Sustainable Return of Investment (SROI) and Green Building Rating System (GBRS) certification. Moreover, the information flow is mapped linking the competences that need to collaborate each other both to provide and gather the information required for the different studies. In particular, the continuous lines refer to the relationships necessary to perform the environmental and life cycle studies, while the dashed lines refer to the workflow for pursuing GBRS certification. Nevertheless, note that the relationships could be direct, i.e. calling for a physical interaction between actors, but even indirect, i.e. simply requiring the sharing of specific design documents and/or data. In addition, it is worth mentioning that to give evidence of the active role played starting from the design process by clients and producers, for falling under OmniClass classification, they are represented in the graphs by the facility use disciplines and the construction disciplines respectively. Indeed, producers are directly linked with contractor and construction company, whereas the facility use disciplines stand for the clients, since in some cases they turn out to be also the owner of buildings.

12.2 The Diversity of Networks Tied to Project Complexities

For surveying current state of design firms, the information flow analysis is conducted on a sample of building design projects, developed in different design practice. Indeed, based on the type of practice (e.g. integrated or exclusively architectural) and the level of complexity of the projects usually deal with, the aim is to explore and figure out different way of practice and distinguishing features. In particular, three design practice are examined in term of information flows required to perform the environmental and life cycle studies.

The first practice is an Architectural and Engineering (A/E) practice of about 70 employees located in San Francisco. It handles projects with very different levels of complexities (Fig. 12.2, projects P1-h and P2-h), as represented by P1-h, that is very complex due to the inherent features, the federal requirements and the goal to achieve high LEED certification, and by P2-h, distinguished from a design point of view by a low level of complexity, since not aligned with federal mandates and GBRS certification. The dissimilarity between the two projects is visible starting

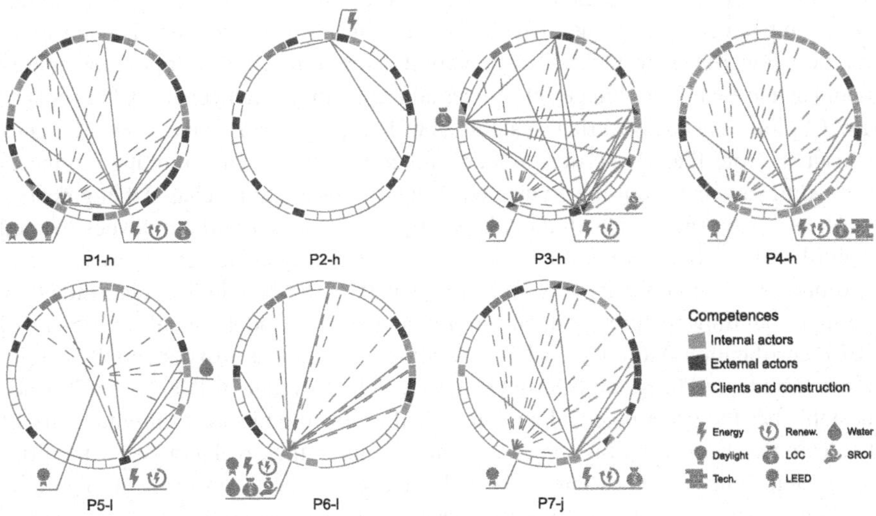

Fig. 12.2 Environmental information flow comparison of different building projects

from the design team directory, composed of a wide range of competences in P1-h as opposite to P2-h where they are significantly reduced and assigned mainly to external parties. The same discrepancy occurs also with regard to the environmental and life cycle studies performed and the connected information flows demanded. In P1-h the actors in charge are environmental design/sustainability and building energy design, responsible, on one hand, for LEED checking, water and daylight studies and, on the other, for studies on energy, renewables and LCC respectively. Instead, since in P2-h there are not environmental specialists, the same architects are committed to perform energy simulations, as unique type of environmental study planned for the project. Concerning collaboration, P1-h is characterized by a thick network of relationships, in which building energy design and environmental design/sustainability constitute the cornerstone. To run studies on energy, renewables and LCC, the collaboration is required with both architectural and engineering competences (Zou and Alam 2020). Architects are involved to identify building shape and orientation; total area of floor, wall and window; types of glazing, wall and roof as well as the related energy efficiency measures. Whilst mechanical, electrical and plumbing engineering are engaged to define the proposed systems for their field of expertise, including renewables systems and hints for the integration of energy efficiency measures. In addition, cost estimation is consulted for first cost, operation and maintenance cost, replacement cost and total building energy cost, claiming in some cases the interaction with producers to set the life expectancy of systems. Moreover, based on building targets set by clients, building energy design cooperates with environmental design/sustainability, to gather data about water consumption for sizing solar hot water panels and to provide data for building energy performance and on-site renewable energy credits. Indeed, environmental design/sustainability is responsible for water and daylight studies as well as for pursuing LEED certification and the collection of related documents, calling for the interface with the specific reference competences (see Chap. 10). Conversely, in P2-h the network and information flow towards environmental sustainability is minimized, developing the energy model on the basis of architectural choices and consulting engineers for what concern HVAC, plumbing and electrical systems as well as plugs loads and energy efficiency measures. Note that in the design team directory the lack of plumbing engineers implied the attribution of the connected competences to the mechanical engineers.

The second practice is still an A/E practice but of about 200 employees located in Washington, DC. Here building projects are generally characterised by high and medium level of complexity (Fig. 12.2, projects P3-h, P4-h and P5-l), determined by the main building features, federal requirements (not mandatory only in P5-l) and LEED certification. Moreover, in P3-h the complexity increases exponentially in relation to the adopted Integrated Design Bid Build delivery methods (Kereri 2017), since it establishes the synergic development and, after few months even simultaneous, of the design and construction processes. This results in the involvement for the principal competences of both internally and externally experts, covering architecture, mechanical engineering, electrical engineering, environmental design/sustainability and A/E management. Regarding environmental and life cycle studies, they affected the different design process according to projects complexity, encompassing studies on energy, renewables, LCC and LEED certification, with the addition of SROI for P3-h and studies on technologies for P4-h. However, they diverge in terms of responsible actors, shifting from the two pivotal competences, represented by environmental design/sustainability and building energy design, to multiple expertise, by adding as in the case of P3-h the so-called life cycle cost modeller for LCC and the internal department of decision economics for SROI. In particular, LCC studies are carried out to comply with federal requirements (Fuller 2010), accounting both the architectural part with wall, roof and to a limited extent also glazing and slab, and the engineering part, including renewable systems and if relevant HVAC and electrical systems. To check the cost/effectiveness, for each technological element are evaluated the following costs: initial, maintenance, major repair, replacement, demanded energy and (only for renewables) generated energy. To go further, SROI studies consider from a triple bottom line perspective the benefits derived from the solutions of glazing, roof, renewables systems and building efficiency measures for energy and water. In this way, it is confirmed the key role of the responsible actors within the decision-making process and the relevance of their choices, here expressed in terms of energy bill savings, water bill savings, greenhouse gas savings and air pollutants savings. Finally, studies on technologies focus on constructive solutions to reach Energy Star Roof, evaluating various materials, components and green roof, taking into consideration the operations and maintenance costs as well as the content of recycled roofing materials (Blackhurst et al. 2010). Overall, the combination between, on one hand, the environmental and life cycle studies performed and, on the other, the range of actors involved establishes the network and information flow demanded to the different projects, appearing very complex and intricate in P3-h and less dense and complicated in P5-l.

The third practice is an Architectural practice of about 60 employees located in Chicago and distinguished by building projects with good level of complexity (Fig. 12.2, projects P6-l and P7-j). They involve almost the same range of competences, assembled according to the peculiar type of building and assigned half internally and half externally the design firm in question. Nevertheless, note that since the practice represents an architectural branch of an integrated firm established worldwide, when engineers turn out to be internal, they refer to actors placed in the most various offices. During design, studies on energy, renewables, LCC and LEED certification are developed, attaching as appropriate water and SROI studies. Concerning the network, in P6-l the information flow gravitated towards a single actor who exploited the information collected by all other subjects to develop the most varied investigations on the project. Instead, in P7-j the network of relationships revolved around the two main responsible actors, identified in: the building energy design and the environmental design/sustainability. Deepening the environmental and life cycle studies and the connected information flow, in both cases the actors in charge cooperated with the same competences, constituted especially by architecture, mechanical, electrical, plumbing, cost estimation and clients. However, surveying different technological elements, they required the data collection about the specific items considered, interacting in a more or less intense way with the connected reference actors.

12.3 Switching from One-to-One to Shared Communication

From the information flow analysis of the sample of building design projects, the distinguishing features of the different practice come to light, emphasising some hotspots for improving in the near future the development of environmental and life cycle studies. Shared goals are primarily the enhancement of the relationships among architects, mechanical, electrical and plumbing engineers and the strengthen of interaction with building energy design and the environmental design/sustainability (Keeler and Vaidya 2016).

The A/E practice accustomed to design both particularly complex projects and simpler projects demands the establishment of very different information flows along the process towards environmentally sustainability. On the matter, it is important to note that the participation of environmental experts is excluded in the less complex projects. Here only energy simulations are performed directly by the architectural team, attesting how they are trained to the use of energy modelling tool, although in a simplified way if compared to specialists. This proves the importance of spreading environmental issues as much as possible within the design team to promote a greater awareness of the design solutions adopted towards environmental sustainability and, especially, life cycle perspective.

Instead, when A/E practice are used to deal with projects characterized by high complexity, dictated in particular by federal mandates, the working environment is different. In this case, the office position turns out to be strategical since encourages the collaboration with central government and thus the assignment of the design of very challenging service facilities due to the stringent requirements and objectives imposed, including the mandatory development of LCC studies. For facilitating the co-design among the engaged competences, the practice is composed by both architectural and engineering disciplines to promote integrated planning and thus optimize time and costs as well as the same design process. Moreover, it embeds in-house all main competences of each discipline, promoting the overall exchange of information between actors, even when some of them rely formally on external consultants or internal subjects located in different offices. Indeed, the information flow increases exponentially when multiple actors are in charge of environmental and life cycle studies, calling for a greater communication, shared data and efficient information management aimed at optimizing the workflow.

Narrowing the scope to Architectural practice, it emerges the constantly need to collaborate and interact with engineering disciplines to cope with and meet environmental and life cycle issues (the same occurs in reverse order). In this way, to fill the engineering gap, the design team depends on external actors under potentially different design firms. The advantage to allocate them to people of the same design firm in question but located in different office is that they know and share the same way of practice, being accustomed to work together and in synergy. Indeed, this statement lapses in case of cooperation with external parties, in which there is a need of a progressive definition of the information flow and relationship demanded, unless external partners are recurring and so mostly aligned each other.

Regardless of the type of design practice and projects complexity, the entire network of relationships is expected to change in the near future for improving process efficiency. Indeed, to date the information flow occurs via one-to-one relationships and through the most various supporting documents, calling for a continuous interaction over the process and, in some case, the accounting of outdated data. To fill the gap, the creation of one-single shared record (Lai et al. 2019) containing the entire set of life cycle information will allow practitioners to save time in finding data and to easily access to the most updated data for the widest design purposes.

References

Baldwin AN, Austin SA, Hassan TM, Thorpe A (1999) Modelling information flow during the conceptual and schematic stages of building design. Constr Manag Econ 17(2):155–167

Blackhurst M, Hendrickson C, Matthews HS (2010) Cost-effectiveness of green roofs. J Archit Eng 16(4):136–143

Fuller S (2010) Life-cycle cost analysis (LCCA). National Institute of Building Sciences, An Authoritative Source of Innovative Solutions for the Built Environment, 1090

Keeler M, Vaidya P (2016) Fundamentals of integrated design for sustainable building. Wiley

Kereri JO (2017) A comparison of project party relationships in design-bid-build and design-build delivery methods. Int J Archit Eng Constr 6(4):26–32

Lai H, Deng X, Chang TYP (2019) BIM-based platform for collaborative building design and project management. J Comput Civil Eng 33(3):05019001

OmniClass (2012) A strategy for classifying the built environment. Table 33—Disciplines

Zou PX, Alam M (2020) Closing the building energy performance gap through component level analysis and stakeholder collaborations. Energy Build 224:110276

Chapter 13
Worksheets to Implement Life Cycle AE(C) Practice by Sharing Responsibilities

Abstract To switch from current design practice to life cycle AE(C) practice, a change management is needed to support during design the establishment of a life cycle-oriented decision-making process. Due to the wide range of information demanded, the sharing of responsibilities plays a key role, as evidenced by the personal worksheets intended to assign to each competence the committed technological elements as well as the connected life cycle information. In particular, the involved competences are: architecture, interior design, health/laboratory design, structural, mechanical, plumbing and electrical engineering, building energy design and environmental design/sustainability. They are all invited to follow the progressive implementation of information aiming at a growing level of detail and accuracy of data over the process. The purpose is twofold. On one hand, the worksheets specify the life cycle criteria to be implemented for encouraging the team in adopting a systemic approach to life cycle design. On the other, their fulfilment allows to collect the specifications of each technological element to be clearly communicated to contractor.

13.1 Detailing the Assessment Framework to Suit Design Process

Based on the in-depth analysis of current design practice (see Chaps. 8, 9, 10, 11, 12), the challenge is to envision the change management of A/E firms in order to implement Life Cycle Thinking (LCT) into design process. The goal is reached interrelating theory and practice. On one hand, theory is represented by the knowhow developed in the field of research, deepening LCT and related methodologies. On the other, practice is represented by the experience in the field, deepening the design processes of a sample of building design projects. Life cycle information and design process are thus correlated on the base of current design practice with the aim to explore LCT for decision-making support. The effort is to bridge the gap between theory and practice in order to endorse design firms in the transformation, identifying the necessary key shifting both in thinking and in process in line with environmental targets and life cycle perspective.

A. Dalla Valle, *Change Management Towards Life Cycle AE(C) Practice*,
PoliMI SpringerBriefs, https://doi.org/10.1007/978-3-030-69981-9_13

The integration of LCT into design process is gained staring from the proposed assessment framework (see Chap. 6), assuming its application ex-ante in order to truly orient the decision-making. To this end, the framework is further developed, adjusted and tailored to fit the most demanding projects, with reference to new healthcare facilities committed to federal mandates and LEED certification. Contrary to the original basic matrix of the framework, the updated version is based, on one hand, by the different stages of life cycle from cradle to grave and, on the other, by the different phases of design process. In this way, building process (UNEP 2014) is narrowed to design phase, excluding construction, in use and end-of-life phases and deepening design process according to standard classification (AIA 2014). Concept phase is breakdown in Pre-Design (PD) and Schematic Design (SD), while design phase is split in Design Development (DD) and Construction Documents (CD).

For steering decision-making in line with life cycle perspective, the life cycle information to keep in mind during design process are outlined according to the minimum submission requirements defined by federal mandates (Department of Veterans Affairs 2010), in order to follow as much as possible the current advancement of design tasks. In fact, to integrate LCT in practice in a feasible and actionable way, the framework aims to support and promote along the process, on one hand, a progressive implementation and, on the other, a gradual detailing of life cycle information according to the process development (Hollberg et al. 2020). For this purpose, the set of life cycle information is specified in agreement with the main technological elements commonly faced during design process. In particular, for architectural disciplines, they refer to: cladding, envelope, walls, floors, finishes, equipment, furnishing. While for engineering disciplines, they concern: structure, HVAC, plumbing, electrical and renewables systems. Note that structure corresponds to load-bearing building structures; walls and floors to internal partitions; and finishes to just the superficial layer of interior walls, ceiling and flooring. Cladding involves only the external siding of envelope, while envelope includes the whole set of layers of the external wall, roof and slab on grade, as intended for building energy simulations. Lastly, equipment stands for the series of technical tools and appliances; and furnishing consists of home furniture and accessories.

With reference to the established elements, the framework firstly defines the technological elements (e.g. cladding) to be considered for each life cycle topic (e.g. construction materials) involved of each life cycle stage (e.g. product stage). In some cases, all technological elements of both architecture and engineering are embraced, such happens for transport issues. In others, they are only partially included, since the life cycle topics affect only certain building elements, such happens for maintenance operations. The resulting life cycle information associated with the technological elements (for cladding e.g. construction materials, transport to site, etc.) are mapped within the framework in relation to design process, distinguishing between current practice (in blue) and life cycle-oriented practice (in green). Current practice is represented by the combination of the life cycle information accounted during the decision-making process of the sample of analysed building design projects (see Chap. 11). Whereas life cycle-oriented practice is conceived, envisioning the implementation of LCT into design process starting from the available information and depicting the connected life cycle topics to be included in the different phases of the process. For instance, concerning finishes, it is possible to include in the decision-making process not only aesthetic and economic issues but also the related VOC emissions. Likewise, in the comparison of construction materials, the design team can embrace maintenance and replacement issues as well as end-of-life issue within the set of adopted criteria.

In this way, the design team is expected to integrate within the framework the recommended life cycle information according to the design phase in progress. Moreover, for boosting the optimization of the process, the whole set of life cycle quantitative data turns out to be the thresholds to be not exceeded in the subsequent phases of the process, if not expressly justified and validate. This applies to the transition from one phase to another of the design process as well as to the shifting from design to construction process. Indeed, in a life cycle practice, the final life cycle information and data are included in the construction specifications, placing constraints on the selection process of the bidding phase. This is why, despite focusing only on design process, the book calls for life cycle AE(C) firm, to stress the impact of A/E team choices on the whole process, especially on the construction phase.

The key factor is that, for supporting the change management demanded to shift from an A/E firm to a life cycle AE(C) firm, the framework set out the life cycle information to be considered during the design process by sharing responsibilities. Due to the amount of required information, to put the framework into practice, it is in fact pivotal to shares the roles among design actors and make them responsible for gathering the assigned life cycle information. In particular, the involved competences turn out to be: architecture, interior design, health/laboratory design, structural engineering, mechanical engineering, plumbing engineering, electrical engineering, building energy design and environmental design/sustainability.

13.2 Supporting Architectural Disciplines in Life Cycle Design

Envisioning a life cycle-oriented practice, design team is in charge not only, as happens nowadays, of the design of the committed technical building elements but also of finding and evaluating along the process the connected life cycle information. The appointed actors are therefore supported by the framework in accounting the life cycle information during the decision-making of each phase of the process. Moreover, to not overburden their assignments, since each competence is entrusted with one or more technological elements and all connected life cycle information, the framework endorses a progressive implementation of the information and a growing level of detail and accuracy over the process.

To encourage the team in adopting a systemic approach for life cycle design, a personal worksheet has been inferred from the framework for each responsible competence, explaining the main life cycle information at issue for each technological element appointed in relation to the design phase. The purpose is twofold. On one hand, they disclose the life cycle criteria to be implemented within the process to orient the decision-making process in line with life cycle perspective. On the other, they turn out to be the specifications to be embedded for each technological element within the BIM model/database (assimilable to the Level of Development) and to be clearly communicated to contractor. Below a brief explanation of the worksheets related to architectural disciplines.

With reference to the proposed assessment framework, architecture team is in charge of the design especially of cladding, envelope, walls and floors solutions, being thus entrusted with the connected life cycle information as further life cycle-oriented criteria for the decision-making process (Fig. 13.1). In PD phase, the focus is mainly on building massing, pushing the selection of the best building scheme not only in relation to architectural, functional and environmental issues (e.g. orientation, view and energy consumption) but also on the amount of construction materials (Konis et al. 2016). Due to the very preliminary phase and the non-definition of materials, envelope area is taken as reference data, since meaningful for massing comparison. In SD phase, architecture effort involves more technological elements, partitioning envelope in glazing and massive parts and start working on roof and cladding solutions. In addition, depending on the different functional areas, walls and floors begin to take shape, derived firstly by circulation flow and blocking

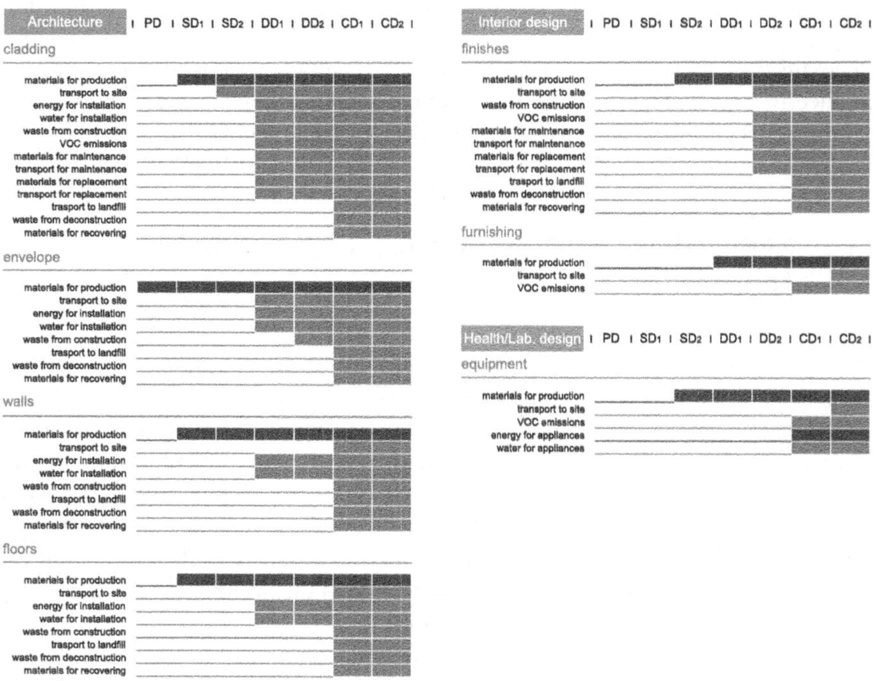

Fig. 13.1 Worksheets to support architectural disciplines in life cycle design

and secondly by rooms and common areas. In the evaluation of different alternatives, the surface is taken as reference data for glazing, wall and floor. Instead, since the team has to propose to client different cladding solutions (Brookes and Grech 2013), beyond performance metrics and aesthetic features, materials quantity is taken as further assessment criteria as well as the distance from manufacturing site for reducing transport impacts. In DD phase, architects give details to the assigned technological elements, gradually defining the quantity of construction materials. Indeed, cladding and envelope are here specified in the first case in terms of panels size, while in the second in all technological layers. By contrast, internal partitions are set out gradually, starting from the definition of the typical wall and floor solutions to end with the layering of all included technological packages. Moreover, to orient the decision-making process of cladding and envelope solutions, beyond materials quantity, the transport distance from factory to building site, the amount of construction waste as well as VOC emissions (Loftness et al. 2007), materials and transport for maintenance and replacement processes are established as further life cycle-oriented criteria (Madureira et al. 2017). Instead, in overall terms, architecture team has to consider the expected energy use for products installation, assuming for instance a literature percentage based on construction materials (Hong et al. 2014), and the water use for on-site production, depending on the adopted wet/dry solutions (Allen and Iano 2019). Finally, in CD phase, architects review the previous claimed life cycle information, extending transport and end-of-life specifications to all assigned technological elements, including waste derived from deconstruction process and materials for reuse, recycling and energy recovery (Ng and Chau 2015). All details are clearly communicated to contractor as minimum requirements for the bidding phase, in conjunction with the complete drawings fully coordinated with other disciplines.

During the process, interior design team is responsible for the development of finishes and furnishing solutions, being thus committed of gathering and considering the related life cycle information (Fig. 13.1). In SD phase, they start working on finishes, proposing different types of materials, but it is only in DD phase that they are detailed before at room level and after to the whole building, defining the set of furnishing required for each space (Binggeli 2008). In the choice of finishes materials, they are invited to take into account not only technical performance and aesthetics but also the following life cycle criteria: transport distance, use emission and maintenance and replacement operations (Aktas and Bilec 2012). In CD phase, interior design team finalizes both finishes and furnishing, updating the declared life cycle information and adding the amount of waste and the end-of-life scenarios for finishes, whilst the transport distance and VOC emissions for furnishing (Jaakkola and Quansah 2010). All information is clearly communicated to contractor as minimum requirements for the bidding phase, combined with materials samples and colour palettes.

Besides the traditional architectural and interior competences, specialized competences are required for the design of certain building types, as in the case of healthcare facilities and research laboratory buildings. They provide guidance about the functional and organisational setup of floors and they are in charge for the selection of the necessary building equipment, being entrusted with the related life cycle information (Fig. 13.1). In SD phase, health/laboratory team supports architects in the arrangement of functional blocks, defining the internal organization of each functional area and proposing the major equipment for providing the service. In DD phase, they move forward with the selection of appliances and accessories before for typical rooms and after at building scale, simply accounting the number of items for each equipment. Lastly, in CD phase, when the team opts for the specific set of building equipment, life cycle criteria are implemented during the decision-making (Aksamija 2010), including transport distance, use emissions, energy and eventually water consumption, striking a balance with functional and operational aspects (Lopez et al. 2017). All details are thereafter considered as minimum requirements for each type of equipment.

13.3 Extending Life Cycle Design to Engineers and Specialists

As happens for the architectural disciplines, also engineering disciplines and specialists are involved in the change management for shifting from an A/E firm to a life cycle AE(C) firm. For implementing a life cycle-oriented practice, each responsible team has to design the assigned technological elements taking into consideration during the decision-making the life cycle information stated by the framework. The progressive collection of such information and data allows, on one hand, the overall monitoring of the process and, on the other, the verification or not of its optimization. In particular, the actors in charge turn out to be structural, mechanical, plumbing and electrical engineers and, for what concern specialists, building energy design and environmental design/sustainability.

During design process, structural engineering team is in charge of the development of building structure and of the connected life cycle information for orienting the decision-making process in line with life cycle perspective (Fig. 13.2). In SD phase, they usually evaluate different structural systems for the selected architectural scheme (De Wolf 2014), specifying the set and amount of materials for each alternative and including deliberations on transport issues, to start implementing life cycle criteria. In DD phase, the team is called to detail the selected system in order to integrate into the architectural and mechanical assets. Here, besides reviewing the previously accounted life cycle information, it has to declare the expected amount of waste (if any), distinguishing for instance the different share of prefabricated elements and on-site constructions (Jin et al. 2020). In CD phase, the structural system is finalized, calling for the adjustment and/or confirmation of the amount of materials by type, distance and means of transportation and quantity of waste produced (Webster 2007). They are all clearly communicated to contractor, attached to the complete structural systems drawings fully coordinated with other disciplines and suitable for bidding.

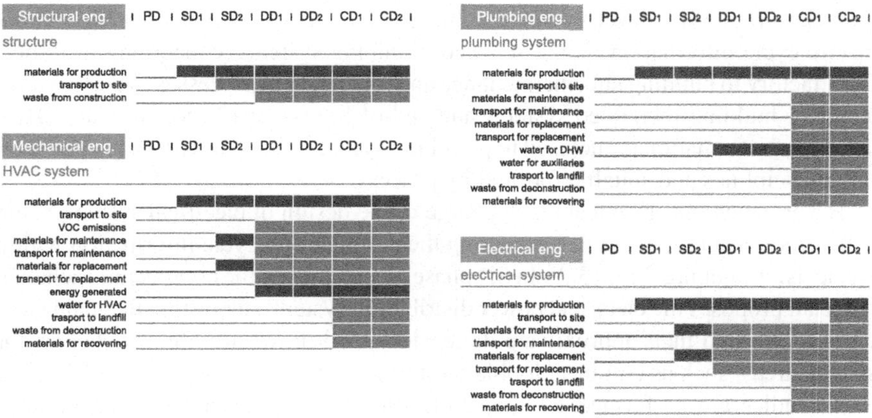

Fig. 13.2 Worksheets to support engineering disciplines in life cycle design

Mechanical engineering team is responsible for the development of HVAC systems solutions, being thus committed of gathering and considering the related life cycle information within the process (Fig. 13.2). In SD phase, after surveying existing underground utilities, they assess different types of HVAC systems with the related energy conservation measures, locating and sizing the mechanical equipment rooms. To select the best system alternative, LCC studies are developed for each cooling and heating option (Wang et al. 2020), including maintenance and replacement process. In DD phase, HVAC floor plans are elaborated for typical areas, setting the main air distribution and piping layouts. In this context, the team is called to design minimising the linear distribution of conduits in order to reduce the amount of materials, estimating the energy generated. In CD phase, mechanical engineers are committed to detail the planned HVAC systems, updating the previous life cycle information on system materials, maintenance activities, replacement processes and energy generated by non-renewable resources. Moreover, they are encouraged to add further information related to transport, water used for operation as well as to end-of-life issues, including transport to landfill, waste derived from deconstruction process and potential materials for reuse, recycling and energy recovery. All these specifications are clearly communicated to contractor as minimum requirements for the bidding phase as well as the fully coordinated HVAC systems drawings.

The design of plumbing systems is committed to plumbing engineers that result therefore in charge of the related life cycle information, considered as further criteria for the decision-making (Fig. 13.2). In SD phase, the team focuses on utilities, suggesting approximate size and location of treatment and storage facilities and setting out a preliminary layout to check how they relate to sanitary, storm and water systems. In DD phase, they detail plumbing system plans, specifying equipment and fixtures and calculating water consumption for domestic hot water (Fuentes et al. 2018), to start implementing a life cycle perspective. In CD phase, in accordance with the progressive level of detail achieved, the team is responsible for follow-up life cycle information on the selected plumbing systems, concerning: transport from factory to building site, maintenance and replacement processes, water used for ancillary building-integrated systems and end-of-life issues (Piroozfar et al. 2016). The set of information is enclosed to plumbing systems plans, fully coordinated with other disciplines and suitable for bidding process.

Electrical engineering team is in charge of the design of electrical system, being called for considering and implementing the pertinent life cycle information within the decision-making (Fig. 13.2). In SD phase, depending on the architectural scheme, the team proposes the electrical power distribution system, suggesting the main electrical rooms and their approximate dimensions, and different lighting schemes for special areas, such as entrance, lobbies, waiting rooms and public areas. Moreover, LCC studies are performed for major electrical equipment and lighting systems (Principi and Fioretti 2014), including in the assessment the life cycle information on maintenance and replacement operations. In DD phase, electrical engineers itemize building plans, identifying power supply system, lighting layouts for typical rooms and all electrical closets. In this way, they are responsible for the quantification of system assets and for the progressive information updating about activities and transport for maintenance and replacement processes. In CD phase, based on the final electrical system, they adjust the declared life cycle information, attaching additional information on transport distance from factory to building site and the systems end-of-life, including transport issues and the amount of landfilled materials compared to recovered materials. The whole set of information constitutes the minimum requirements for bidding, integrating fully coordinated electrical drawings.

During design process, building energy design team is specialized in the assessment of renewables systems and building operational energy, being entrusted with the connected life cycle information (Fig. 13.3). In PD phase, it conducts climate analysis and annual energy studies, at least for heating, cooling and lighting, for optimizing building orientation and massing and steering design choices from an environmental point of view. In SD phase, the team starts performing preliminary renewable energy

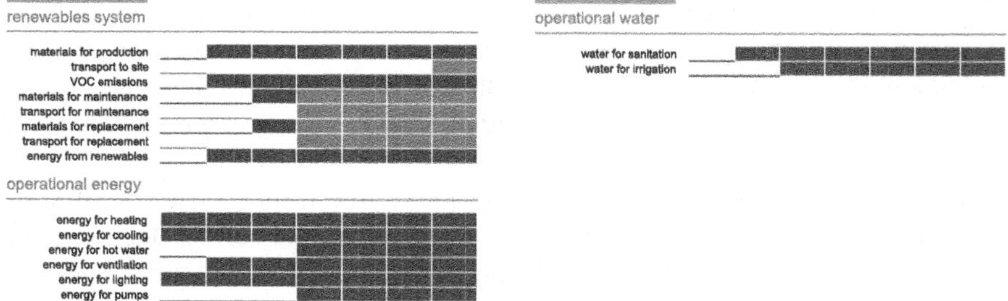

Fig. 13.3 Worksheets to support specialist disciplines in life cycle design

calculations and energy consumption in coordination with mechanical and electrical engineers. Along with the matching of energy generated by renewable resources with the building energy use for heating, cooling, lighting and ventilation systems, it is invited to consider further life cycle information. They concern the selected renewables systems in terms of Green House Gas emissions and maintenance and replacement processes. In DD phase, renewables and operational energy calculations are rerun according to the evolution of both architectural and engineering parts, evaluating several energy efficiency alternatives to achieve project energy reduction goals. Here the simulations cover aslo the energy used for hot water supply system, for pumps and auxiliaries and for miscellaneous equipment. Finally, in CD phase, the team provides the expected operational energy profile of the building (Collinge et al. 2018), breakdown into the whole set of powered systems, and as appropriate information on renewables, including transport, maintenance and replacements issues. Renewables specifications are trasmitted to contractor as minimum requirements for bidding, whereas operational energy profile has to be checked during the use phase, to verify if the realized building operates as designed.

Another key specialist team is environmental design/sustainability, since represents the main coordinator for the achievement of Green Building Rating System (GBRS) certification and the subject responsible for the development of some of the connected studies, among which building water consumption (Fig. 13.3). To this end, staring from the outset of the process, the team looks into GBRS scorecard, indicating the attainable credits and points and opening debate on their implementation. In SD phase, it conducts preliminary water use calculations, in coordination with plumbing and civil engineers, assessing sanitation systems for indoor evaluation and irrigation system for the outdoor ones, disclosing the recycling strategies proposed for reducing water usage for landscaping (Stec and Kordana 2015). In DD and CD phase, a progressive level of detail is achieved for building operational water profile, updating GBRS scorecard according to the final design solutions selected by both architectural and engineering disciplines and collecting all design documents to submit for the review process (Ma and Cheng 2016). As expected for energy issues, operational water has to be verified during the use phase, to ascertain if calculations are close to reality.

Nevertheless, since design processes are unique and project-based, it is worth mentioning that the suggested worksheets inferred from the framework are not valid in absolute terms, because of they refer to the most virtuous life cycle-oriented practice feasible at current state. For this reason, they can be adjusted according to the peculiar features of the project at issue, on one hand, providing some levels of simplification for less complex projects and, on the other, overcoming them in the next future when more life cycle information will be available. Despite this, note that since life cycle information must be intentionally mapped, the personal worksheets represent an important supporting tool for the involved actors, offering them from the outset the complete list of information within their competence to be accounted along the design process.

References

Aksamija A (2010) Analysis and computation: sustainable design in practice. Design Princ Pract Int J 4(4)

Aktas CB, Bilec MM (2012) Service life prediction of residential interior finishes for life cycle assessment. Int J Life Cycle Assess 17(3):362–371

Allen E, Iano J (2019) Fundamentals of building construction: materials and methods. Wiley

Binggeli C (2008) Materials for interior environments. Wiley

Brookes AJ, Grech C (2013) The building envelope: applications of new technology cladding. Butterworth-Heinemann

Collinge WO, Rickenbacker HJ, Landis AE, Thiel CL, Bilec MM (2018) Dynamic life cycle assessments of a conventional green building and a net zero energy building: exploration of static, dynamic, attributional, and consequential electricity grid models. Environ Sci Technol 52(19):11429–11438

De Wolf C (2014) Material quantities in building structures and their environmental impact. Doctoral dissertation, Massachusetts Institute of Technology

Demkin JA, The American Institute of Architects (2014) The architect's handbook of professional practice. Wiley, New Jersey

Department of Veterans Affairs (2010) A/E submission requirements for VA medical center major new facilities, addition and renovation. Program Guide PG 18–15, Volume B

Fuentes E, Arce L, Salom J (2018) A review of domestic hot water consumption profiles for application in systems and buildings energy performance analysis. Renew Sustain Energy Rev 81:1530–1547

Hollberg A, Genova G, Habert G (2020) Evaluation of BIM-based LCA results for building design. Autom Construct 109:102972

Hong T, Ji C, Jang M, Park H (2014) Assessment model for energy consumption and greenhouse gas emissions during building construction. J Manag Eng 30(2):226–235

Jaakkola JJ, Quansah R (2010) Building materials and furnishing. Occup Environ Lung Diseases 69

Jin R, Hong J, Zuo J (2020) Environmental performance of off-site constructed facilities: a critical review. Energy Build 207:109567

Konis K, Gamas A, Kensek K (2016) Passive performance and building form: an optimization framework for early-stage design support. Sol Energy 125:161–179

Loftness V, Hakkinen B, Adan O, Nevalainen A (2007) Elements that contribute to healthy building design. Environ Health Perspect 115(6):965–970

Lopez JB, Jackson D, Gammie A, Badrick T (2017) Reducing the environmental impact of clinical laboratories. Clin Biochem Rev 38(1):3

Ma J, Cheng JC (2016) Data-driven study on the achievement of LEED credits using percentage of average score and association rule analysis. Build Environ 98:121–132

Madureira S, Flores-Colen I, de Brito J, Pereira C (2017) Maintenance planning of facades in current buildings. Construct Build Mater 147:790–802

Ng WY, Chau CK (2015) New life of the building materials-recycle, reuse and recovery. Energy Proc 75:2884–2891

Piroozfar P, Pomponi F, Farr ER (2016) Life cycle assessment of domestic hot water systems: a comparative analysis. Int J Construct Manag 16(2):109–125

Principi P, Fioretti R (2014) A comparative life cycle assessment of luminaires for general lighting for the office–compact fluorescent (CFL) vs Light Emitting Diode (LED)—a case study. J Clean Product 83:96–107

Stec A, Kordana S (2015) Analysis of profitability of rainwater harvesting, gray water recycling and drain water heat recovery systems. Resourc Conserv Recycl 105:84–94

United Nations Environment Programme, Antink R, Garrigan C, Bonetti M, Westaway R (2014) Greening the building supply chain. UNEP Sustain Build Clim Initiative

Wang C, Xu A, Jiao S, Zhou Z, Zhang D, Liu J, Wang Y (2020) Environmental impact assessment of office building heating and cooling sources: a life cycle approach. J Clean Product 121140
Webster MD (2007) Structural design for adaptability and deconstruction: a strategy for closing the materials loop and increasing building value. In: New horizons and better practices

Chapter 14
Progressive Life Cycle Information Flow Within a BIM-Oriented Working Environment

Abstract In view of the change management needed to turn into a life cycle AE(C) practice, it is presented the life cycle information flow, pointing out for each design phase the set of actors directly involved with the committed technological elements and life cycle information. By showing an overall and systemic vision of the process, the aim is to support especially managerial disciplines in the identification of the main design competences that must be strengthened for the accomplishment of life cycle tasks. The joint combination of all actor efforts and connected life cycle information results in a project-based and well-framed set of data of the facility during the whole life cycle, envisioned in a BIM-oriented working environment. The project life cycle database provides a long-term vision of the facility, ensuring an added value for both clients and design firms and a continuous built-up of know-how.

14.1 Vision of a Project Life Cycle Database

As widely known, the integration of Life Cycle Thinking (LCT) into design process and thus the implementation of a life cycle AE(C) practice demand a shift both in thinking and in process. The shift in thinking concerns the understanding that every building needs to be designed as a system keeping in mind the entire life cycle of its elements. Occurring mainly at individual level, this change is supported by the personal worksheets which identify for each competence the corresponding life cycle information to be progressively evaluated for each assigned technological element (see Chap. 13). On the contrary, the shift in process refers to the awareness that the design team itself need to function as a system, where every actor has to grasp how the decisions undertaken by each affect the decision made by all others, with the aim to jointly design and achieve high-performance buildings towards the entire life cycle (Boecker et al. 2009). In this way, the change occurs at collective level for the whole design team, stressing collaboration, cooperation and communication (Shen et al. 2010), calling for and strengthening the role of project management disciplines.

To provide a synopsis of a life cycle-oriented design process, the related information flow is disclosed for each phase of design process, pointing out the set of actors

A. Dalla Valle, *Change Management Towards Life Cycle AE(C) Practice*,
PoliMI SpringerBriefs, https://doi.org/10.1007/978-3-030-69981-9_14

directly involved with the committed technological elements and life cycle information. The objective is twofold. Firstly, the identification of the main design competences to be strengthened or implemented towards life cycle perspective. Secondly, an overall vision of the life cycle information required by each actor at each phase, representing a key driver for both project and building management purposes (Camposano and Smolander 2019). Indeed, from project team point of view, the progressive implementation of the life cycle information is meant as the minimum submission requirements committed at each design phase, for internal use (voluntary path to encourage a life cycle-oriented practice) or external use (mandatory path if required in the near future by building directives). Conversely, from building point of view, the joint combination of actors and assigned life cycle information is intended as the basic frame to develop a project-based and well-framed set of data of the facility during the whole life cycle, envisioned in a BIM-oriented working environment. In fact, following the worksheets recommendations, all actors in charge and/or allowed contribute information to and extract information from the building virtual model, providing a long-term vision of the facility. The resulting life cycle BIM database allows the establishment in one-record of the life cycle information and data of the building at issue, fostering a systemic vision of the project and representing an added value for clients and a continuous built-up of know-how.

The information flow of life cycle AE(C) practice is depicted (Fig. 14.1) taking the same representation adopted for current practice (see Chap. 12), showing the main responsible competences in accordance with OmniClass classification (OmniClass 2012). However, in this instance the responsible actors enter the life cycle information into the BIM model/database of the facility, instead of exchange information as currently happens with one-to-one relationships among the involved competences. In this way, one-single record collects all life cycle information, enabling actors to insert their assigned life cycle information but also to use all available information and data to carry out the most varied design studies. Indeed, they represents the input data required to perform the most miscellaneous environmental and life cycle studies, supporting in particular the inventory phase for Life Cycle Assessment (LCA) (EN 15978:2011) but also Life Cycle Costing (LCC) (ISO 15686-5:2017) studies.

14.2 Development at Early Phases of the Design Process

With the aim to build up the life cycle BIM database of the project, it is crucial to have the support of all main competences, committed to retrieving and entering the life cycle information and data of the assigned technological elements. The joint combination of all individual efforts results thereafter in the overall and systemic vision of the designed building, represented by the life cycle BIM database.

Since the implementation of life cycle practice is a challenging task, demanding a shift both in thinking and in process, an overview of the required change management is provided to bring into alignment the project development with life cycle perspective. By combining the proposed personal worksheets (see Chap. 13), it is thus

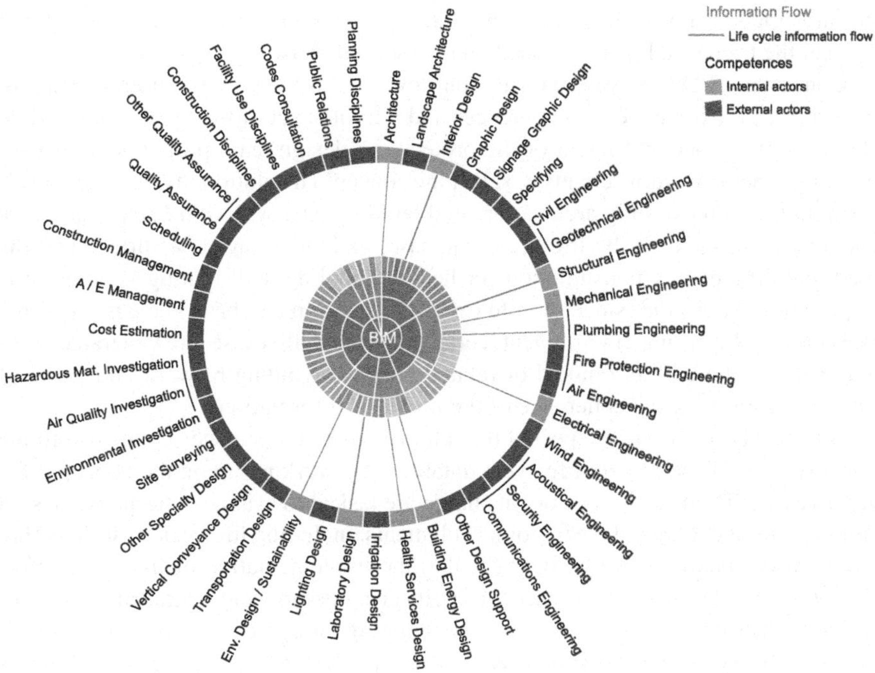

Fig. 14.1 Information flow of life cycle design practice

explained the life cycle information flow recommended for each phase of the process, strategical especially for a managerial viewpoint. Indeed, the project management competences have a pivotal role in the transformation, being called for soliciting the involved actors in the accomplishment of the life cycle tasks and for verifying the consistency of the retrieved life cycle information. In this context, note that during the design process the actors in charge remain almost the same: what changes are the life cycle topics included and the technological elements to be progressively evaluated, starting from the early phases of design process. In particular, they concern Pre-Design (PD) phase and Schematic Design (SD) phase, in which crucial decisions made for the project but with impact on the whole building process (Bueno et al. 2018).

For implementing life cycle perspective into design practice, PD phase represents a crucial step, notably for the definition of the project sustainability goals to be pursued in the subsequent phases. Here, managerial competences are invited to expressing sustainability goals in accordance with the list of life cycle information to encourage their monitoring during the project development. This implies the combination of the targets commonly required by clients and requirements, such as energy use reduction, water use reduction and greenhouse gas emission reduction, with the ones valuable from a life cycle perspective. In this way, they are enlarged for instance with construction waste targets and recycling materials targets, taking as a reference the thresholds set up by Green Building Rating System certification criteria and going beyond the traditional goals focused on operational phase.

Concerning PD life cycle information flow, due to the very preliminary state of the project, it involves few competences and a limited set of life cycle information. Based on the scope and intent of the project, the design team proposes alternative massing scheme to allow clients to select the conceptual solution to be further developed. In this context, only architecture and building energy design competences are encouraged to kick off the life cycle database, including envelope surface and the main building energy consumption for heating, cooling and lighting systems. It is important to stress that since the life cycle information database is envisioned in a BIM-oriented working environment, certain data, as in this case, are generated automatically by the building model or related plug-in, avoiding re-work and waste of time and improving the coherence of the provided information.

During SD phase, it is expected that all principal competences contribute to the improvement of the project life cycle database, by working up on the concept plan approved in PD phase. The actors in charge are tasked to evaluate alternative design solutions for the assigned technological elements, including life cycle criteria within the decision-making process. In particular, the implementation of life cycle information is stressed from the outset for engineering components, requiring mechanical and electrical engineering as well as building energy design to provide specifications on the maintenance and replacement processes of the proposed solutions for HVAC, electrical and renewables systems respectively. Instead, since architectural competences are mainly focused on building shaping, aesthetic and functional issues, only cladding transport is demanded as additional life cycle information. The same occurs for structural engineering with reference to the transfer of load-bearing materials. Moreover, combining the set of preliminary architectural and engineering solutions, the life cycle database is enriched with the operational energy and operational water used for the principal systems. On one side, building energy design has to estimate the energy consumption of heating, air conditioning, lighting and ventilation systems as well as the energy generated by renewables system, with the connected emissions. On the other, environmental design/sustainability has to esteem the water consumption of sanitation and irrigation systems. Furthermore, as design process within BIM-oriented working environment, all technological solutions for both construction material and building systems are progressively modelled and quantified as appropriate by area, volume, item or linear development.

14.3 Completion at Advanced Phases of the Design Process

Although it is crucial to begin implementing LCT starting from the early phases of design process, it is equally strategical to continue applying life cycle perspective at advanced phases and onward (Cavalliere et al. 2019), possibly to the extent of encompassing the whole building process. To this end, the life cycle information flow aimed, on one hand, at steering decision-making towards life cycle design and, on the other, at establishing a project life cycle database, is extended for covering also the most advanced phases of design process (Hollberg et al. 2020). They include Design Development (DD) phase and Construction Document (CD) phase, calling for the full involvement of all leading architectural, engineering and specialist disciplines.

The DD phase is intended to add an increased level of detail for all design aspects to further define interior and exterior spaces, in terms of structure, cladding, envelope, walls, floors, finishes, equipment, furnishing as well as HVAC, plumbing, electrical and renewables systems. Consistently, the responsible teams have to explore the main life cycle topics, related in particular to production, construction and use stages, entering in the life cycle database the information and data of the principal elements. Concerning the construction stage, transport issues and construction waste have to be taken into account by architecture for envelope and cladding and by structural engineering for building structure. In addition, architecture is in charge of quoting the energy used for products assembly and installation and for the water used for on-site production. About the use stage, architecture and interior competences have to include, respectively for cladding and finishes solutions, the related VOC emissions and the materials and transports required for maintenance and replacement operations. The latter have to be updated also for HVAC, electrical and renewables systems by mechanical, electrical engineering and building energy design. Moreover, building energy design has to recalculate, based on the last architectural and engineering solutions, the operational energy consumption by end use, including: heating, air conditioning, hot water supply, ventilation, lighting systems, pumps and auxiliaries and miscellaneous equipment. The operational energy is then compared with the energy generated both by renewable resources (e.g. photovoltaic cells, wind mills, solar thermal panels, ground or air source heat pumps) and non-renewable resources (e.g. gas boilers, combined heat power, fuels cells and heat pumps). Similarly, the environmental design/sustainability has to rerun the operational water consumption breakdown in sanitation, domestic hot water and irrigation systems.

During CD phase, the project is finalized by the design team with the aim to provide the necessary documents for the bidding phase and to proceed with building construction. The whole set of life cycle information is checked, adjusted and updated both in relation to design development and data sources, opting preferably for primary data of specific products rather than database or literature data. Furthermore, additional life cycle information is demanded, enlarging the technological elements to be considered and the life cycle topics in question. In this way, transport issues are solicited not only for structure, cladding and envelope but also for walls, floors, finishes, equipment, furnishing and for all major building systems, involving the responsible competences. The same happens for construction waste, extended to walls, floors and finishes, appointing in particular architecture and interior design. These competences with the support of health/laboratory design have to insert also the use emissions of cladding, finishes, equipment and furnishing. Information on materials and transport for maintenance and replacement processes has to be completed for both the architectural part, including cladding and finishes, and the engineering part of mechanical, plumbing, electrical and renewables systems. Concerning building performance, the expected operational energy split by end use has to be reprocessed by building energy design in collaboration with mechanical engineering. Instead the operational water use has to be further specified by environmental design/sustainability for water sanitation and irrigation systems, by plumbing engineering for domestic hot water and building auxiliaries systems, by mechanical engineering for HVAC systems and by health/laboratory design for the main building appliances. Finally, end-of-life information has to be implemented into the life cycle database involving, on one hand, architecture and interior for all construction materials and, on the other, mechanical, plumbing and electrical engineering for all systems materials. The competences in charge are therefore encouraged to assess the transport to landfill, the waste derived from deconstruction process and the potential materials for reuse, recycling and energy recovery of the assigned technological elements.

Following the proposed life cycle information flow, the life cycle database is progressively built up and completed during the design process. From life cycle perspective, it comprises in one-single record a project-based and well-framed set of data of the facility during the whole life cycle, conceived, in line with current trends, in a BIM-oriented working environment (Chong et al. 2017). In this way, it turns out to be an added value both for clients and design firms, enabling a continuous built-up of know-how and representing the input data to develop the most miscellaneous design studies, especially LCA and LCC (Santos et al. 2019). By contrast, from design process perspective, it allows the overall monitoring of the process as well as the verification or not of its optimization, since the assessed life cycle data stand for the thresholds to be not exceeded in the subsequent phases of the process. They become in fact the set of specifications to be clearly communicated to contractor as minimum requirements for the bidding phase. In this way, the life cycle database developed by sharing roles and responsibility supports the change management actually underway in building sector overcoming environmental sustainability towards life cycle implementation. However, if on one hand, this organizational set-up represents a point of strengths to put LCT in practice, not overburdening design competences in their current assignments; on the other, it constitutes a point of weakness, since it involves a wide range of competences that must be trained and successfully managed to embed life cycle perspective in practice.

Since the integration of LCT into design and building processes represents an upcoming and challenging task for the involved actors (EC-JRC 2020), the implementation of the proposed life cycle information flow is considered as the springboard for a life cycle AE(C) practice, to be advanced by follow-up investigations.

References

Boecker J, Horst S, Keiter T, Lau A, Sheffer M, Toevs B (2009) The integrative design guide to green building. Wiley, New Jersey

Bueno C, Pereira LM, Fabricio MM (2018) Life cycle assessment and environmental-based choices at the early design stages: an application using building information modelling. Arch Eng Design Manag 14(5):332–346

Camposano JC, Smolander K (2019) Inter-organizational integration in the AEC/FM industry. In: International conference on advanced information systems engineering, pp 64–79. Springer, Cham

Cavalliere C, Habert G, Dell'Osso GR, Hollberg A (2019) Continuous BIM-based assessment of embodied environmental impacts throughout the design process. J Clean Product 211:941–952

Chong HY, Lee CY, Wang X (2017) A mixed review of the adoption of Building Information Modelling (BIM) for sustainability. J Clean Product 142:4114–4126

EN 15978:2011. Sustainability of construction works. Assessment of environmental performance of buildings. Calculation method

European Commission—Joint Research Center (2020) Level(s)—a common EUf ramework of core sustainability indicators for office and residential buildings. Publications Office of the European Union: Luxembourg

Hollberg A, Genova G, Habert G (2020) Evaluation of BIM-based LCA results for building design. Autom Construct 109:102972

ISO 15686-5:2017. Buildings and constructed assets. Service life planning. Part 5: Life-cycle costing

OmniClass (2012) A strategy for classifying the built environment. Table 33—Disciplines

Santos R, Costa AA, Silvestre JD, Pyl L (2019) Integration of LCA and LCC analysis within a BIM-based environment. Autom Construct 103:127–149

Shen W, Hao Q, Mak H, Neelamkavil J, Xie H, Dickinson J, Xue H (2010) Systems integration and collaboration in architecture, engineering, construction, and facilities management: a review. Adv Eng Inform 24(2):196–207

Chapter 15
Key Findings, Possible Stakeholders and Outlook for Life Cycle Practice

Abstract One of the key emerging challenge of construction sector is the implementation of life cycle AE(C) practice, aimed at making aware decisions, avoiding shifting problems, gaining long-terms perspectives, optimizing design process, leading decision-making and decreasing construction impacts. The resulting change management calls for, especially in particularly complex projects, a new professional figure that acts as intermediary between clients and project team towards life cycle design. Given the topicality of issues, a wide panel of stakeholders may be interested in the outcomes, covering the expertise area (e.g. AEC firms, academy), the assets area (e.g. service companies, software developers) and the information area (e.g. clients, users). Moreover, improvement areas are highlighted for further enhancing and deepening in the next future the integration of life cycle perspective into building process.

15.1 New Competence in Design and Construction Field

In the ongoing process of transformation and increasingly complexity of building sector, sustainability and environmental issues are gaining even more attention, further emphasizing the change management required to design firms (Renz and Solas 2016). In fact, they call for a holistic and transdisciplinary approach in practice, redesigning the organizational set-ups, the boundaries of knowledge, the roles within the project and the entire supply chain of the built environment. The transformation process involves all actors and assets of building process, including: tangible resources, with the growing digitalization and the spread of tools, and intangible resources, with the increasingly specialization of competences. The problem is that while tangible resources are visible to everyone and everyone talks about them, almost nobody speaks about the so-called invisible technology (Sinopoli 1997), which actually represents the real added value for highly competitive working environments. The interest shifts therefore from the tools used during design to the people and competences engaged, to their organization within the process, to their capacity of

A. Dalla Valle, *Change Management Towards Life Cycle AE(C) Practice*,
PoliMI SpringerBriefs, https://doi.org/10.1007/978-3-030-69981-9_15

collaboration, cooperation and coordination as well as to the communication and information flow demanded to meet environmental issues and life cycle perspective (Rezgui et al. 2010).

For this purpose, combining theory and practice, the book is primarily focused on the deepening of current AEC practice to understand how they are equipping and reorganizing themselves in order to address and meet environmental issues (see Chaps. 5, 8, 9, 10, 11, 12). Afterwards, keeping up on current trends and based on design practice, the core is the field of change management required to implement life cycle perspective into design process, envisioning a life cycle AE(C) practice (see Chaps. 13, 14). It promotes the understanding of building environment in a holistic and systematic way as well as the interaction among different disciplines, focusing on knowledge, competencies, life cycle information and the related information flow required for bringing out the necessary transformation. From a design standpoint, this turns out to be strategical for orienting decision-making towards life cycle design, for monitoring and optimizing design process and for establishing life cycle database of buildings.

The switchover from current practice to life cycle AE(C) practice represents a challenging task for the design team with respect to two key issues. On one hand, it requires a shift in thinking because buildings must be conceived and designed as systems, in which each subsystem and component affects and in relation with the others, considering the whole life cycle. On the other, it demands a shift in process, since the same design team must work like a system, in which all members understand how their choices affect the decisions made by all others. For this reason, especially when dealing with particularly complex projects, a new professional figure is expected to emerge, recalling the "middleware", capable of adopting and promoting a systemic vision of the project and process from an environmental and life cycle perspective (Fig. 15.1).

Indeed, the proposed change management suggests the introduction of a new subject assimilable to a middleware that, as happens in IT sphere, deals with the interrelation between users who requested a service and the basic resources by means of a highly heterogeneous distributed system (Razzaque et al. 2015). This figure acts as an intermediary (Krenz et al. 2014) between the client (user) who entrusted the project (service) and the actors of the process (basic resources), with the task of activate networks and information flows (distributed systems) among the different parties involved (heterogeneous). Nevertheless, a crucial aspect is the approach and the way in which it provides the service required: on one side, it adopts and supports a systemic vision of the project and, on the other, it fosters the process optimization from an environmental and life cycle perspective, leading, monitoring and managing the invisible technology. In this way, the new competence, demanded in particular for highly complex projects, is involved starting from the early brief to identify in accordance with clients the sustainability goals and targets required for the project and to be thereafter constantly checked during the process. Moreover, since each design process is unique and project-based, he is committed to adjust and if needed simplify the set of life cycle information in relation to project complexity and requirements as well as to define the environmental and life cycle studies to be performed

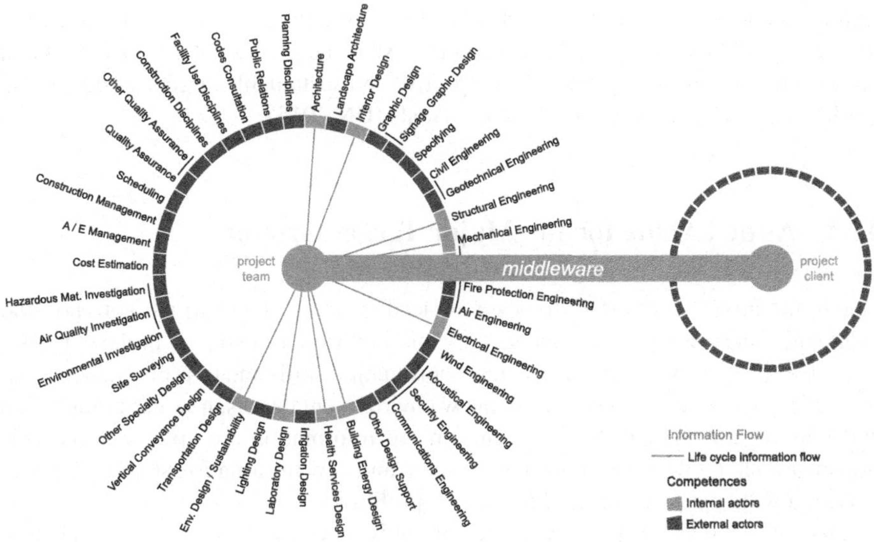

Fig. 15.1 New professional figure assimilable to a middleware

during the process. Based on this, he updates the personal worksheets, identifying the main design competences and relationships to be strengthened or implemented to meet project goals, stressing and supporting the involved actors for considering and collecting the assigned life cycle information and data. Indeed, he is responsible for encouraging the spread of life cycle criteria during the decision-making and for the progressive collection of life cycle information during the process, fulfilling the project life cycle database with a growing level of detail to improve the reliability of the accounted data. Another key task is the check of data consistency between the ones declared by different actors and the ones included into design studies, in order to encourage the communication, collaboration, coordination and information flow among the engaged design disciplines as well as with producers, contractors and clients. Finally, depending on the asserted life cycle information and developed design studies, he is in charge of tracking the optimization or non-optimization of the design process conducted for the proposed building. In this way, he supports the project development and, at the same time, the change management of the process, converting resources first into "capabilities", i.e. minimum capacities, and then into "maturity", i.e. quality achieved by the integration of good practices.

The inclusion of the new professional figure assimilable to a middleware proves not only the need for new competences in the construction field, but also how the question of interdisciplinarity becomes in some cases neo-disciplinarity. If the suggested life cycle AE(C) practice is put in action, significant reverberations would be visible at different scales. Firstly, the adoption of Life Cycle Thinking (LCT) in the built environment with the resulting change of mindset of the actors engaged in the design process. Secondly, the implementation of life cycle information and related methodologies as effective decision supporting tool in design process, as claimed in the

European context by Level(s) (EC-JRC 2020). Third, the activation in the construction sector of the type of mechanisms able to start the process of enhancement and optimization in line with life cycle perspective and sustainable targets, as demanded worldwide by Sustainable Development Goals (UN 2015).

15.2 Added Value for the Major Target Groups

Due to the broad spectrum of topics embraced in line with construction trends and challenges, the book provides an added value in different fields: (i) in the expertise area identifying new competences, new coordination skills and thus new educational path; (ii) in the assets area suggesting new environmental tools, new communication and new organizational tools; (iii) and in the information area pointing out new organizational models, new data management and new information sharing. For this reason, a wide panel of stakeholders might profit of it.

First of all, the A/E practice object of study, because they have the chance to observe and examine in a critical way their design process and workability. In fact, being engaged as sample of building case studies, they obtain an overview of the design process of the selected projects, grasping their environmental efforts during the decision-making process and their way of practice. Moreover, they can figure out how to turn into a life cycle AE(C) practice, integrating LCT into design process and changing management in order to optimize and streamline the decision-making process in line with environmental targets and life cycle perspective (Rezaei et al. 2019).

In addition, all AEC firms can be interested in the results because throughout the surveying of case studies and the suggested life cycle AE(C) practice, they can be aware of reference practice and forwards models, encompassing the related strategies and methods. Indeed, despite the potential transformations vary in relation to firm size, some improvement can be shared for instance about the overall decision-making, the optimization of design process and the plan of action to implement life cycle perspective in practice.

In second place, the whole construction industry followed by the associated service companies and software developers may be concerned, because they can discover points of weakness and strength looking over current practice and innovation systems for the proposed life cycle AE(C) practice. In this way, they can launch advanced tools and develop new methods to improve and optimise, on one hand, design process and the connected information flow and, on the other, the environmental performance of buildings and construction supply chain (Meex et al. 2018). In addition, new kinds of business with new types of integrated services and new professional skill profiles might arise to suit the application of life cycle perspective in practice. In this regard, it is important to note the advantages obtained by design firms if the proposed graphical representations as well as suggested project life cycle database are converted into digital supporting tools, speeding up the process and the impact on projects, especially in a BIM-oriented working environment.

Moreover, also regulators and legislators can have a stake in the book outcomes because throughout the overview of AEC firms they become aware of how policies, regulations and standards affect design practice and the resulting change management. This can lead to edit and adjust the set of laws, rules and building codes that are in effect, to improve environmental performance of building sector, to foster some environmental issues rather than others and to implement LCT and related methodologies (in particular Life Cycle Assessment and Life Cycle Costing) as common practice (Säynäjoki et al. 2017). In this perspective, for instance, the personal worksheets can be embedded into project requirements, making the progressive evaluation and implementation of life cycle information mandatory for the responsible design team along the decision-making process.

Furthermore, also clients and users can take advantage from the book becoming informed about current trends and design practice towards environmental sustainability. Note that this is crucial for the whole construction sector, since they actually represent one of the most important drivers for sustainable building (Kilinc et al. 2015) and their awareness is pivotal to lead the transformation process.

Lastly, another stakeholder that can be interested in the results is the academic community. Indeed, the overview on current A/E practice and the envisioned life cycle AE(C) practice can be intended in support of the adjustment of academic programmes (Fullan 2006) in order to bring up both architects and engineers capable to design sustainable buildings. To this end, for instance, it is important to encourage students in taking a life cycle perspective during decision-making, promoting in the educational paths multidisciplinary learning activities and information exchange among different disciplines.

15.3 Future Prospects for Life Cycle AEC Practice

Given the wide range and relevance of topics dealt with, in the next future the work could be deepened in many parts and in relation to several topics.

The exploration of the proposed assessment framework, personal worksheets and life cycle database can be extended to the other phases of building process (Antink et al. 2014), including not only A/E firms but also Construction companies and related actors in the change management of their practice towards environmental sustainability and life cycle perspective.

Moreover, the technical and technological implementation of the proposed project life cycle database within BIM could definitely spread LCT in building practice. In this perspective, the first step is the insertion of life cycle information into BIM, enriching the set of information just embedded in the model and connecting when possible the data with the relative parametric objects (Durão et al. 2019). The second step is the grouping of information according to the phase of the design process, including a wider range of data with the advancement of the process. The third step is the insertion for each life cycle information of the additional linked data, such as the responsible actors and source used in the data collection.

In addition, to disseminate life cycle AE(C) practice, follow-up investigations are offered. On one hand, they concern the sources now available for retrieving the life cycle information of the main technological elements, in order to assist the actors in charge in entering and completing the assigned tasks. On the other, they regard the life cycle information actually taken into account within the different life cycle tools (Al-Ghamdi and Bilec 2017), in order to encourage practitioners in taking advantage of the established life cycle database. Both explorations are solicited in relation to the different phases of design process, to promote LCA studies from the early design phases (Dupuis et al. 2017).

Finally, the effective application of the suggested life cycle AE(C) practice starting from the beginning of the project could represent a turning point in order to assess the impact on the decision-making and the change management of design process. To this end, a pilot project could be proposed and build-up, to truly verify the optimization of design process and, consequently, of buildings in line with sustainable goals and life cycle perspective. In this way, the life cycle information and data included in the project life cycle database would become, according to the different phases of the process, pivotal reference values for future construction projects and processes.

References

Al-Ghamdi SG, Bilec MM (2017) Green building rating systems and whole-building life cycle assessment: comparative study of the existing assessment tools. Journal of Architectural Engineering 23(1):04016015

Antink R, Garrigan, C, Bonetti M, Westaway R (2014) Greening the Building Supply Chain. UN Environment Programme

Dupuis M, April A, Lesage P, Forgues D (2017) Method to enable LCA analysis through each level of development of a BIM model. Procedia engineering 196:857–863

Durão V, Costa AA, Silvestre JD, Mateus R, Brito J (2019) Integration of environmental life cycle information in BIM objects according with the level of development

European Commission – Joint Research Center (2020) Level(s) – A common EUf ramework of core sustainability indicators for office and residential buildings. Publications Office of the European Union, Luxembourg

Fullan M (2006) The future of educational change: System thinkers in action. J Educ Change 7(3):113–122

Kilinc N, Ozturk GB, Yitmen I (2015) The changing role of the client in driving innovation for design-build projects: stakeholders' perspective. Procedia Economics and Finance 21:279–287

Krenz P, Basmer S, Buxbaum-Conradi S, Redlich T, Wulfsberg JP (2014) Knowledge management in value creation networks: establishing a new business model through the role of a knowledge-intermediary. Procedia CIRP 16:38–43

Meex E, Hollberg A, Knapen E, Hildebrand L, Verbeeck G (2018) Requirements for applying LCA-based environmental impact assessment tools in the early stages of building design. Build Environ 133:228–236

Razzaque MA, Milojevic-Jevric M, Palade A, Clarke S (2015) Middleware for internet of things: a survey. IEEE Internet of things journal 3(1):70–95

Renz A, Solas MZ (2016) Shaping the future of construction: A breakthrough in mindset and technology. World Economic Forum

Rezaei F, Bulle C, Lesage P (2019) Integrating building information modeling and life cycle assessment in the early and detailed building design stages. Build Environ 153:158–167

Rezgui Y, Hopfe CJ, Vorakulpipat C (2010) Generations of knowledge management in the architecture, engineering and construction industry: An evolutionary perspective. Adv Eng Inform 24(2):219–228

Säynäjoki A, Heinonen J, Junnila S, Horvath A (2017) Can life-cycle assessment produce reliable policy guidelines in the building sector? Environmental Research Letters 12(1):013001

Sinopoli N (1997) La tecnologia invisibile. Milano, Franco Angeli

United Nations (2015) Transforming Our World: The 2030 Agenda for Sustainable Development

The manufacturer's authorised representative in the EU is Springer Nature Customer Service Centre GmbH, Europaplatz 3, 69115 Heidelberg, Germany. If you have any concerns regarding our products, please contact ProductSafety@springernature.com

Printed and bound by CPI Group (UK) Ltd, Croydon, CR0 4YY
28/09/2026
02224076-0014